THE HANDBOOK FOR STAR TRACKERS

Making and Using Star Tracking Camera Platforms

****

THE HANDBOOK FOR STAR TRACKERS

Making and Using Star Tracking Camera Platforms

by

JIM BALLARD

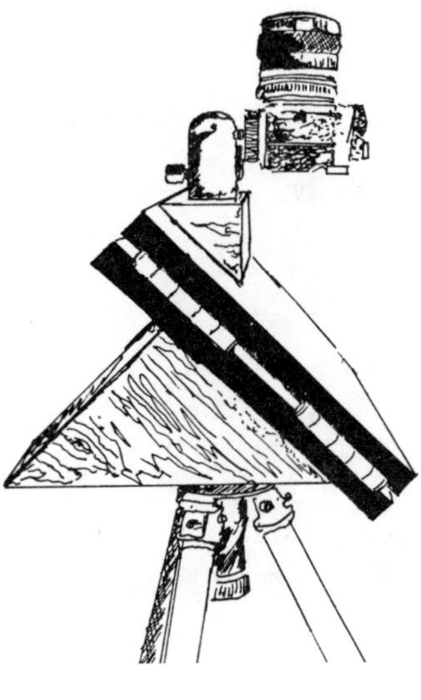

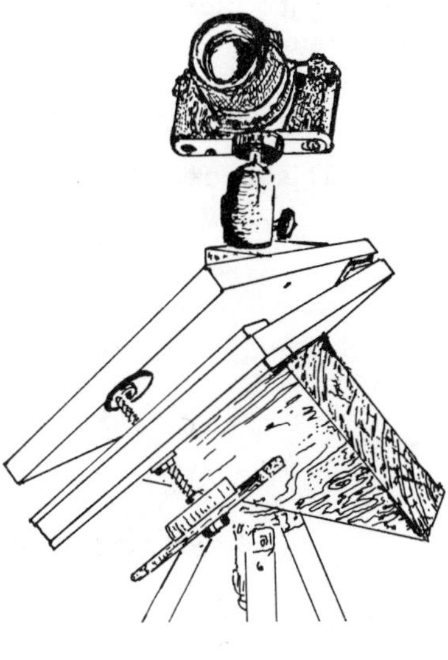

I'd like to thank Robert Stahl and Mark Terry for patiently reading and editing many drafts of this book. Thanks also to Frank Zullo, Barry Gordon, Michael Covington and Dennis di Cicco for their valuable letters and comments about star tracking. William Spaulding, Sr. created an experimental prototype motor control unit that keeps my automated trackers running flawlessly. I appreciate your time and help with this project, Bill.

Photos are by the author, unless otherwise indicated.

Figure 1-1 through Figure 1-4 are reproduced from SKYSHOOTING by R. Newton Mayall and Margaret W. Mayall, Copyright 1968 by Dover Publications, Inc. Used with permission.

Illustrations on page 3 are by Mickey Matsuzawa.

This book is for Laura and Kelly.

Printed in U.S.A.

Sky Publishing Corporation
49 Bay State Road
Cambridge, Mass. 02138

ISBN 0-933346-47-6

TABLE OF CONTENTS

FOREWORD

I suspect that every amateur astronomer who owns a camera has given some thought to capturing the beauty of the heavens on film. Recent technological advances in photographic emulsions have, in many ways, made astrophotography a simple endeavor. With a camera mounted on a tripod, a few seconds' exposure (a mere snapshot compared to what was needed a few years ago) will record all the stars that can be seen with the naked eye. Longer exposures show beautifully colored star trails arching across the sky.

What is the next step up the astrophotography ladder? For many it is a hinge star tracker like the ones Jim Ballard describes in this book. A few simple parts and an evening's labor will produce a device that stops the sky's motion from the camera's point of view. Now an exposure of just 3 or 4 minutes can record stars and nebulae too faint to be seen even with binoculars.

Ballard has paused on this step of the ladder longer than others. He has found a niche very much to his liking, as you can tell by reading his chapter "In the Clearing." He has more experience with hinge trackers than any astrophotographer I've met -- experience that is valuable to people who have built a tracker as well as those contemplating one for the first time. If you are just starting out, look over the assembly chapter before reading the section on parts. This way you'll have a better idea of what's involved.

When I set about making my first "barn-door" tracker, as I call it, I found all the parts I needed scattered around the workshop. But to complete a project without spending money didn't seem right, somehow. In a quick trip to the hardware store, I found a $3 hinge that was much better than the one in my junk box. The tracker was a success, and saying it cost $3 impresses people more than if it cost nothing!

The point of all this is that trackers can be assembled for less than the cost of a few rolls of film. It would seem rather foolish, then, to pay $15 or $20 for a book explaining how to do it. So to keep the cost of this book low, the text has been reproduced directly from the author's typescript.

Dennis di Cicco
SKY & TELESCOPE

PREFACE

"You'd be surprised how many people write and tell me that they want to know how to put together a star tracking platform from scratch. My drawer is full of such letters."

Dennis Harper, owner of Astro Kits, told me that in a recent phone conversation. It seems that a lot of people not only want to assemble kits -- like the one Astro Kits provides -- or buy an assembled, ready-to-track mount, but want to scout out all the materials from beginning to end. This makes sense. It follows the long tradition of amateur telescope makers tinkering here or there to improve their scopes' performance. It gives those do-it-yourselfers great pride to show off an astrophoto made from a self-assembled mount. As it clearly should.

If you are one of those do-it-yourselfers, then this book is for you. This material literally tells you where to buy the nuts and bolts, how to put the parts together, how to test the final mount, and how to take astrophotos with it that will rival -- for unguided shots -- most commercially available units.

I've tried to make the text easy to read, easy to understand, and complete enough so that you can assemble the trackers without any problems. I've described parts that are commonly found in most large hardware stores. Perhaps best of all, these trackers are inexpensive, the simplest one costing less than $15 -- not counting the tripod head or camera gear -- to put together.

You should find the section on dry testing the mounts useful. By photographing a flashlight through various lenses and tracking conditions, you can tell if your star tracker will have any mechanical problems before you turn the unit skyward.

Another section will help your on-site, deep-in-the-night shooting efforts. I will describe how you can use an illuminated guidescope to see if your tracker is precisely tracking the stars.

You will find out how to get your unit exactly polar aligned by using finderscopes and other kinds of sighters.

Other sections will describe how to automate your star trackers, citing specific motors that can be used and where they can be purchased. I've included wiring diagrams for simple and fairly complex motor control units. This section should appeal to those who already have a tracker and want to make it work even better. One section describes commercially available trackers along with approximate prices. Sometimes a commercial mount offers exactly what you need, and the design is too complicated or too expensive to build from scratch. But it's just right to buy. I have a hunch that many astrophotographers have more than one star tracker. Right? ("Well, I needed this one because it has a built-in polar alignment scope and ..."). I know the lines.

This book is not a philosophical book. Nor is it a book that catalogues nebulae, star clusters, constellations and other deep-sky objects. Bookstore shelves are filled with excellent books on these subjects. This is not a book for people who want to piggyback their camera on a telescope. Buy ASTROPHOTOGRAPHY by Barry Gordon for that information. This is not a book for people who want to make their own telescopes. Buy BUILD YOUR OWN TELESCOPE by Richard Berry for this. This is not a book that covers most general aspects of astrophotography. Buy ASTROPHOTOGRAPHY FOR THE AMATEUR by Michael Covington for this.

To my knowledge, no one has pooled together the useful bits and fascinating pieces about making and using star tracking platforms. This book has that information.

It's common to read that photographing stars from atop a tripod or star tracking mount is a fine way of being "introduced" to astronomy. I'd like to think that this kind of astrophotography is an end in itself, a gratifying, exciting activity. A star tracker has its own special place alongside telescopes, radio telescopes and other instruments of discovery.

PRINCIPLES AND DESIGNS

WHAT ARE STAR TRACKING CAMERA PLATFORMS?

If you've ever taken a long time exposure of the starry sky, you know that the movement of the earth will cause star trails across the film. As the shutter remains open, the star light slides across the frame. From a fixed tripod, only short exposures with fast film can reduce or eliminate such star trailing.

To keep the star images sharp over time, the camera must move at precisely the same rate that the earth is revolving, but IN THE OPPOSITE DIRECTION. If this motion is maintained for a number of minutes, stars never seen with the unaided eye can be recorded. Due to some films' emulsion sensitivity, faint star clusters and nebulae can be captured in the kind of detail seen only through telescopes.

The camera must move at a precise rate around an axis or hinge point. That axis must remain perfectly parallel to the imaginary line of the earth's axis during each exposure.

Star tracking platforms perform both these tasks. Once aligned parallel to the earth's axis -- in other words, aligned to the true celestial pole, and when moving at the correct speed -- a star tracking mount can become a sensitive, precision tool. That's why astronomers and photographers have for a number of years drawn on their imagination and ingenuity to develop these instruments.

Star trackers can be hand driven or automated. Excellent results can be achieved using either method. However, it is difficult to match the performance of a finely tuned automated drive. Vibration, timing, and smoothness of turning are more carefully controlled with automated drives than with hand operated units.

Many star tracking platforms and mounts are lightweight, portable and will fit atop an ordinary camera tripod. They can be polar aligned easily and are quite stable during the exposures. You don't need extensive knowledge of either astronomy or photography to use them skillfully. They are merely tools that can open up the night sky in ways you've never seen.

STAR POSITIONS

You must understand two terms in order to fix the position of a star. These are right ascension (R.A.) and declination (dec.). Right ascension is the celestial equivalent of longitude on earth. It starts at the place where the sun moves northwards across the celestial equator, a point known as the Vernal Equinox. Right ascension is measured eastwards across the sky, ranging from 0 to 24 hours. Each of these hours is divided into 60 minutes, and each minute into 60 seconds.

Declination is the celestial equivalent of latitude on earth. With a celestial equator at 0 degrees, declination positions can range northward +90 degrees to the north celestial pole or southward -90 degrees to the south celestial pole.

By using this system, you can quickly locate stars or other deep-sky objects. You can find both a star's right ascension and declination from any star atlas.

When you first begin to photograph the heavens, it may not seem important to know a star's declination or right ascension position. But, as you become more experienced and graduate to longer telephoto lenses and longer exposure times, you will quickly discover how a star's location can influence your results.

Since most photography from a tracker is done with lenses ranging from wide angle to medium telephoto, locating your subject by declination and right ascension often won't be necessary. Usually your target subject can be framed in the camera by knowing how many degrees it lies -- and in what direction it lies -- from a known star or constellation. However, the right ascension and declination system is necessary for locating deep-sky objects, faint nebulae and other subjects with high powered optics and large telescopes.

HOW ARE TRACKING PLATFORMS ALIGNED?

In order to maintain precise star tracking with a camera, a platform must first be aligned to the north or south celestial pole. Once aligned, it can be moved about this axis point at a continuous rate.

The motion about this polar axis becomes the tracker's right ascension (R.A.); the position of the camera upon this axis is the platform's declination (dec.) setting.

This type of design is called an EQUATORIAL MOUNT because of the north/south polar alignment principle. Many of today's star trackers and telescopes use the equatorial principle for locating stars and guiding. The principle, unlike the technology, hasn't become dated.

The following illustrations, reprinted from R. Newton Mayall's and Margaret L. Mayall's 1949 book SKYSHOOTING, show how an early bellows camera was used for photographing the stars forty years ago. The sketches depict how a board supporting the camera was rigidly fixed to the polar axis. A guiding handle allowed the camera to be moved about the polar axis. This design, although primitive compared to today's high-tech devices, gave "reasonably satisfactory results" when used with a guide scope.

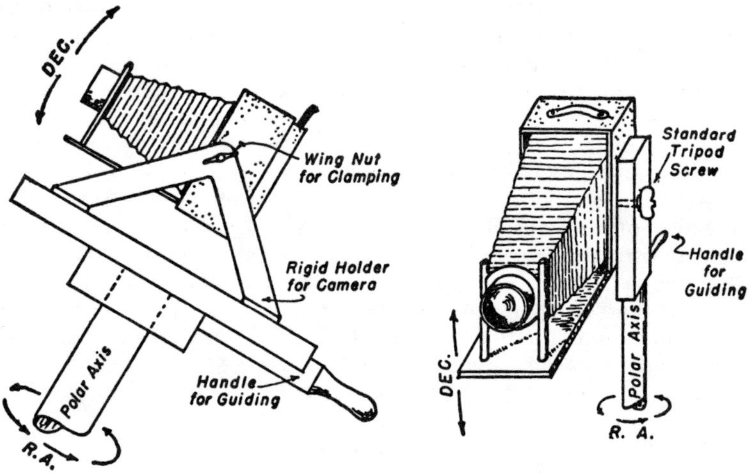

Figure 1-1. A simple equatorial mount showing how a bellows camera was attached for limited guiding.

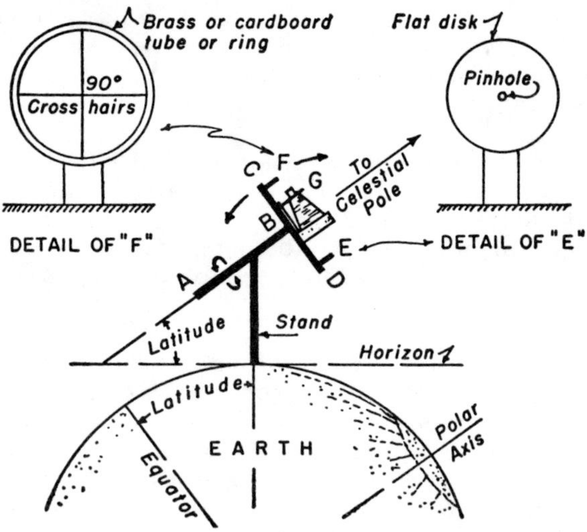

Figure 1-2. Equatorial Mount with one way of polar aligning a platform. A-B rod must point to the celestial pole; C-D platform can be guided by using E-F.

AN EARLY GEAR-DRIVEN CAMERA PLATFORM

One of the earliest gear-driven camera mounts, designed and built by William von Arx, used a rotating metal rod to support the camera platform. The rod, about 1/2 inch in diameter, was positioned between two wooden boards at an angle equal to the photographer's latitude. The upper end of the rod was secured in a bearing; the lower end was housed in a worm gear assembly that could be hand turned. This entire assembly was rigidly mounted atop a metal pipe.

Figure 1-3.

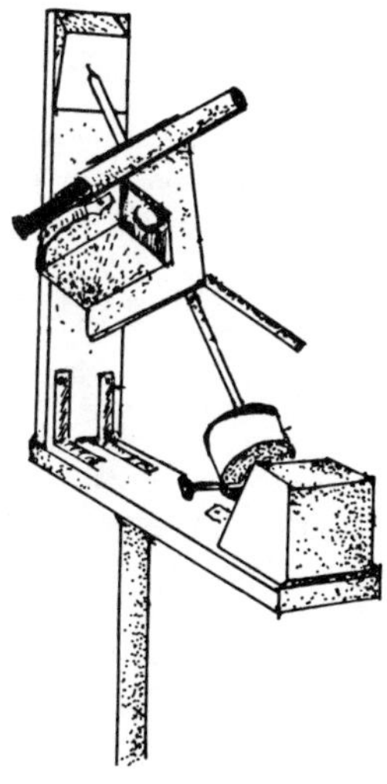

Figure 1-4.

Figure 1-3. Reading glass camera. The simple mounting supports a small telescope and a homemade box camera fitted with a reading glass for a lens. Hand-driven mounting.

Figure 1-4. A Brownie and telescope, mounted atop a platform, is counterbalanced with a weight. Sketches by Mickey Matsuzawa based on photos in SKYSHOOTING.

From the side, the mount looked like a triangle: two wooden sides 90 degrees to one another; the other side -- the platform-holding rod -- the hypotenuse. The camera mount was anchored in the center of the long rod. A finderscope, used for guiding on a star and controlling the rotation speed, was screwed down next to a Kodak box camera.

The mount was aligned by sighting Polaris along the metal rod. Then the photographer could control the rod's turning speed with a thumbscrew attached to the worm gear. If heavier cameras were used, counterweights were attached to another rod perpendicular to the main rod and camera platform.

Twenty-minute non-trailing star shots could be taken with this equatorial mount and gear-controlled platform.

The camera base was mounted directly onto the tracker's wooden platform. Newer ball-and-socket tripod mounts offer more compositional choices and are easier to use when focusing on a specific constellation or section of sky.

SAME PRINCIPLE, MORE COMPACT MOUNTS

Many, if not most, of the current amateur star tracking mounts make use of the equatorial principle. But modern star trackers are more compact, lightweight and portable. Much of the credit for this change must be given to G.Y. Haig, a Scotsman.

In the April, 1975, Observer's Page column of SKY & TELESCOPE, Mr. Haig illustrated and described how he made an equatorial mount from a couple of wooden boards, two hinges, a shelf bracket and some screws, nuts and bolts -- "humble materials," as he put it.

In Haig's design, a hinge line took the place of the metal rod used in von Arx's earlier trackers, greatly reducing weight and size and increasing the tracker's portability. A straight bolt turning through the lower leaf of the platform pushed against the upper leaf and opened it at a precise rate. This "speed control" replaced the need for a more complicated worm gear.

In a letter to me, Mr. Haig explained that he built the drive "in order to try some experiments in stellar spectroscopy, for which I needed exposures of only five or ten minutes at the most; the hinged-board arrangement just seemed to be the simplest way of doing this without becoming involved in more elaborate construction since I had no suitable equatorial mounting available at the time."

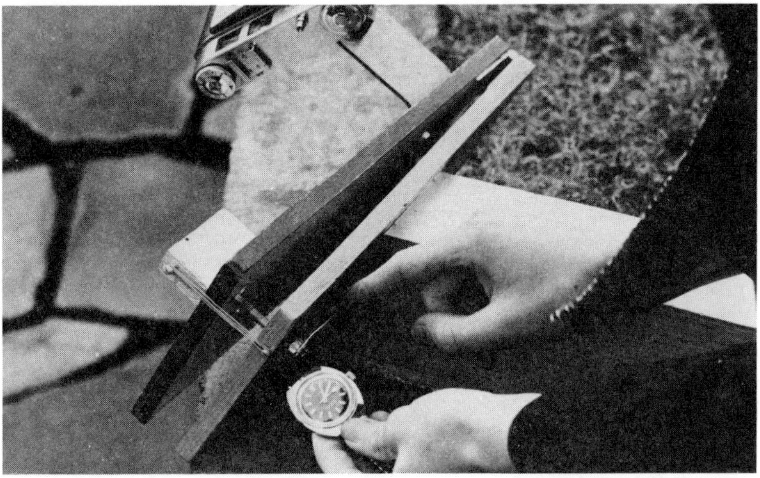

Figure 1-5. This star tracker (often called the Scotch mount) used a straight bolt that pushed directly on the plywood board holding the camera bracket, against the tension of a rubber band. The support block was cut at an angle approximately equal to the latitude of the observing site. Photo by G. Y. Haig.

"I have always felt that an important feature of the mounting," Mr. Haig writes, "was its rigidity when placed directly on a concrete slab or stonework, and was not surprised to read later in SKY & TELESCOPE (e.g. October, 1977) that some amateurs found it necessary to add heavy weights to provide damping for tripod-mounted versions. Polar alignment of the 'grounded' hinge equatorial was easily accomplished with the aid of a mirror."

This kind of mount incorporated a tangent drive. The following discussion details not only why a tangent drive works, but which inherent principles must be considered if you build this type of drive.

THE TANGENT DRIVE PRINCIPLE

To more completely understand how star trackers can be made to move at a precise rate, we must step back and examine the tangent drive and its underlying principle.

A tangent drive is a threaded rod that turns inside a nut, binder post or threaded aperture attached to the base of a mounting and is operated by hand or a motor. One end of the rod bears against a plate (e.g., hinge surface) that rotates about the polar axis (e.g., hinge knuckle). In the diagram below, let X be the point where the tangent arm bears against the plate, and let B be the distance between the polar axis and X.

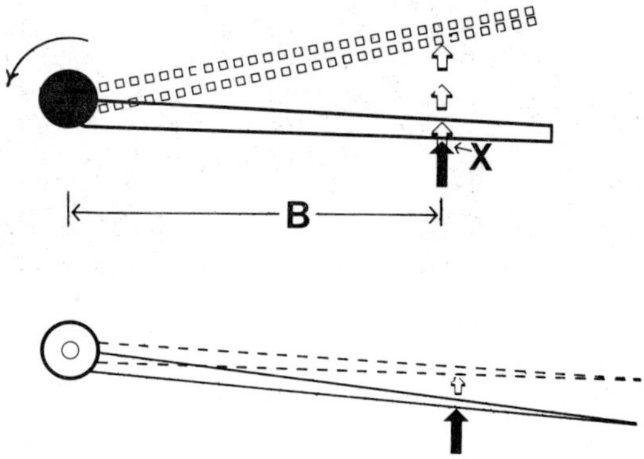

Figure 1-6. Tangent arm principle.

Assuming the rod is 90 degrees to the plate at the beginning of the operation, then the screw must advance by B/57.3 in order to turn the polar axis through an angle of 1 degree.

The earth's sidereal rate is 23 hours 56 minutes, or 86,160 seconds. Dividing 86,160 seconds by 360 (total degrees per revolution) equals 239.3 seconds, the time required to move through 1 degree.

A bit more calculation will show that 0.004375 times B is equal to the required screw advance per minute. This, (finally!) is used to determine the exact distance from the polar axis to where the rod must bear to achieve (at least in the beginning of the operation) the sidereal rate.

As the time increases, a tangent drive loses its rotational speed: The angle at X decreases as the length of the rod increases, and so the tangent decreases. In order to compensate for this effect, the rod may be turned faster.

Or, a curved-bolt rod can be used so that B remains constant through time. The curved rod, as Dennis di Cicco points out in his October, 1985, SKY & TELESCOPE column, becomes a segment of a large worm gear. There is no cumulative error as with a straight-bolt drive.

POSITIONING THE DRIVING ROD

By using the formula:

$$D = 228.6/t.p.i.$$

where D is the distance from the hinge axis and t.p.i. is the number of threads-per-inch of the turning rod, you can quickly calculate the measurements needed for any tangent arm camera platform.

For example, say you have a rod with 32 t.p.i. Dividing 228.6 by 32 gives 7.14, the distance in inches from the center of the hinge knuckle to where the turning rod must bear in order for it to move the plate at the sidereal rate at 1 r.p.m. A few of the most common thread sizes along with the required distance from the polar axis (hinge knuckle center) are given in this table. You can see that the size of your platform will in part determine the size of the rod's threads-per-inch.

MEASUREMENTS FOR DRIVE RODS OPERATING AT 1 R.P.M.

Bolt's Threads-per-inch	Distance From Hinge Center
16	14.28 inches
18	12.70 "
20	11.43 "
24	9.52 "
32	7.14 "

Table 1-1. The bolt must bear against the plate at the precise distance from hinge axis for sidereal rate.

BASIC PARTS AND MATERIALS

Before you begin to build a hinge type tracker, you'll need to mull over the numerous choices of materials and parts. Which size turning screw do you need? What style of hinge? Which finderscope?

I know the joy of rummaging about hardware stores trying to find just the right part to make a project work beautifully. There's no denying that kind of satisfaction. However, sometimes how-to information in books is too sketchy. The builder may discover his or her search for the right part more frustrating than rewarding.

With that in mind, this section is geared more for the weekend do-it-yourselfer than for the seasoned pro who has a milling machine, two lathes and the savvy to build anything at all.

I've included the names and addresses of suppliers of certain parts that are especially useful for constructing a tracker. This may speed up your search for the required materials. Write for availability of parts and current prices.

POLAR SIGHTING DEVICES

A sighting device aligns the axis of the drive's hinge to a true celestial pole. This aid is essential for precision star tracking.

How closely you align the tracker to a celestial pole is one key element in determining how long you can expose the film before unwanted star trailing sets in.

Longer telephoto lenses will require more precise polar sighting devices than normal or wide-angle lenses exposing a subject for the same time. How much alignment error is acceptable is an important subject covered later.

Some polar sighting devices are:

1) The edge of the hinge itself (if the hinge line is small)
2) A straight rod bent into an "open sight" position
3) Hollow items like metal tubing, straws, brackets
4) Telescope finderscopes (with either straight or right angle viewing)

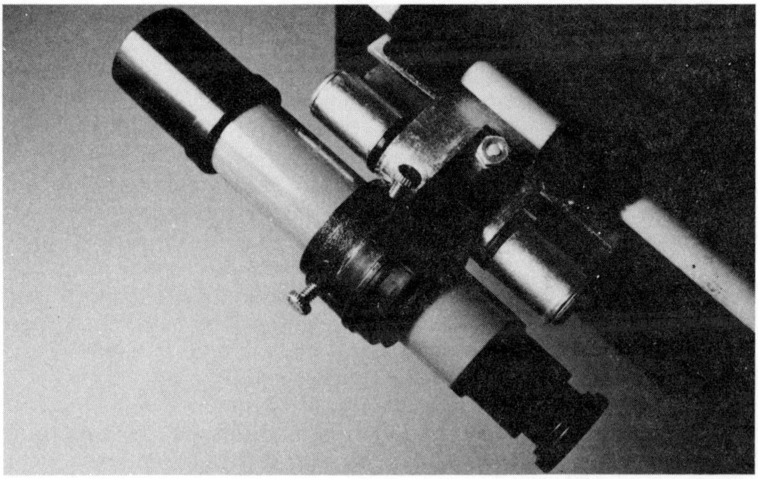

Figure 2-1. A finderscope and open tube sighter are mounted alongside and parallel to the hinge line.

Let's examine each of these aligment aids in detail.

A HINGE may work if its diameter is small enough. The kind used on the Astro Kit platforms narrows to a ridge that's only about 1 mm in diameter, a small enough size to let you sight down its edge (either on the outside or inside) with fairly good accuracy.

Not as precise as a crosshair in an optical finderscope, the hinge, nevertheless, can work well, within the limits of normal and wide-angle lenses. Longer telephotos and extended (fifteen minutes or more) time exposures may require more precise alignment aids.

A large diameter hinge will not work well as a finder. There is too much room for error. You will find it hard to know exactly where the hinge is "pointing." If the hinge has a diameter larger than 1/4 inch, use an additional sighting aid. The extra sighter will pay off in better alignment and, consequently, rounder star images with long exposures.

A STRAIGHT ROD bent up at the ends to form a kind of peep sight works surprisingly well. Use a section of a wire coat hanger or, better yet, buy a toilet flusher replacement rod. This copper rod comes with one end already formed into a circle; just bend equal amounts of both ends 90 degrees to the rod. When done correctly, the height of the tip at one end will equal the height from the rod to the center of the circle. Put a dab of phosphorescent paint on each end of the rod to help frame the pole star. This device is accurate, fast, and easy to make.

TUBE-LIKE sighters are extremely useful because once you know their diameter and length, you can calculate how many degrees you are seeing through the opening.

Here's a formula to remember when figuring how many degrees your tube sighter will show:

A degree is the apparent size of any object whose distance is 57.3 times its diameter.

Astrophotographer Frank Zullo used this information to set up a polar sighting device on his star tracker. In the October, 1985, issue of SKY & TELESCOPE he writes, "A small sighting device attached to the upper board helps align the hinge axis with the north celestial pole, which is located about 0.8 degrees from Polaris. Near the southern hinge a threaded brass tube (a common lamp part at hardware stores) is held by a small angle bracket. An 11/32-inch hole is drilled in another bracket placed exactly 12 inches from the eye end of the brass tube. The hole thus has an apparent radius of 0.8 degrees."

Figure 2-2. Frank Zullo's star tracker uses an open sight for polar alignment. Photo by Frank Zullo.

A ready made tube sighter having about the same diameter as Zullo's is the toilet or vanity water-line supply (Figure 2-1). This gray plastic tube is 12 inches long with an inside diameter of 1/4 inch. Although a bit smaller in diameter, it will nicely fill the bill as a precision sighter.

Another astrophotographer, Ben Mayer, uses a 1/4-inch diameter drinking straw 6 inches long for aligning the STELASCOPE, his custom designed star tracker. He writes, "A plain paper drinking straw aligns the system with the Earth's axis by sighting in the North Star."

Whichever type of open-ended sighter you use -- drinking straw, metal brackets, or plastic tube -- you can get an accurate idea of how many degrees, or parts of a degree, it covers by viewing the full moon through it. The full moon has an apparent size of 1/2 degree. Sight through the finder and see how much of the moon is shown, or how many "moons" it takes to fill the opening. You can even do this measurement during the day if the moon is visible.

FINDERSCOPES will give you exacting control over polar alignment. Included in this group are inexpensive air rifle, BB gun, and .22 rifle scopes as well as the numerous low-power to high-power optical finders for telescopes (Figure 2-1). One quickly apparent advantage to these scopes is their crosshair accuracy. This precision cannot be approached with the devices mentioned earlier.

Another advantage of a finderscope over the other sighting aids comes with its ease of alignment to the hinge's axis. You need to turn only a few thumb screws on the finder's bracket to adjust the scopes; the other devices require more effort to align.

Many finders render the image upside down and left to right reversed. You will quickly get used to this re-orientation, but at first it may be disconcerting. Trying to move 0.8 degrees from Polaris in the correct direction while seeing everything topsy-turvy takes some getting used to! However, once you've got the hang of it, you will be able to get within 1/4 degree of perfect polar alignment consistently.

As with open ended tube-like sighters, you will need to know how many degrees of sky the scope covers. How many degrees are you viewing? Information that comes with the scopes should indicate the viewing coverage, but you may still need to test this using either the moon or the known apparent number of degrees between two stars.

Inexpensive air rifle scopes make good finders. You see everything right side up and left to right as it really is, making it easy to place the crosshairs on a celestial pole.

Inexpensive air rifle scopes have smaller diameter front or back optics than specially designed guide-scopes, thus reducing the brightness of star light reaching your eye. This would be important if you were going to use the scope for tracking during an exposure, but since it is being used as a one-time alignment device, the dimness won't be a problem.

Figure 2-3. An inexpensive 4 X 15 rifle scope mounted parallel to the hinge knuckle axis.

The best, brightest, and most optically sharp rifle scopes can often be purchased at great savings from pawn shops and second hand stores. Some of these hand-me-downs are in perfect condition but can be picked up for as little as one-fourth the original price.

A right-angle attachment can be used with a finderscope to help with the polar alignment. This extension, however, might get in the way when the camera is angled back and over the hinge axis. Try it out and see if you have enough clearance; if so, the right angle attachment may make the task of polar alignment a bit easier.

Peep sights, open tubes, rifle scopes and finder-scopes will each allow you to polar align a tracker with varying degrees of accuracy. Experiment with these aids until you find the best one for your needs.

HINGES

PIN HINGES

Most hinge drives use pin type door hinges to join the upper and lower leaves of the platform. Usually two 3-inch hinges are screwed into 1/2-inch or 3/4-inch plywood or furniture grade board. If you make such a drive, be sure that the hinges are tight without binding, and have no "play" when you open and close them. Too much sloppiness in the hinges can create star tracking problems, especially with longer telephotos and heavier camera bodies.

Using two separate door hinges may give you more lateral stability across the surface of upper leaf than a single hinge. In choosing hinges, or a single hinge, you may need to make a tradeoff between stability and lightness. However, if you intend to photograph the heavens using only normal or wide-angle lenses, either single- or double-pin hinges will work.

STRAP HINGES

You can make an excellent platform with a single, carefully chosen strap hinge. If you decide to try your hand at making a single strap hinge star tracker, find a 10-inch STANLEY (R) hinge that has black nylon spacers about the knuckle. These spacers should be snug enough to eliminate play but still allow smooth movement. This hinge (as well as a hinge without the spacers) carries a model number of V902. Check to see where you can buy it individually, rather than in a package of two. Swing the leaves back and forth to be sure of what you're getting.

If all of the hinges you try have too much play, buy the tightest one. Go to an auto supply store or large hardware and get a 5/8-inch "E" clip. File or sand down one side of the clip to make it just a bit thinner. Carefully hammer the ring onto the hinge knuckle next to one of the spacers. This will eliminate any play but still allow the upper leaf to open smoothly. This size "E" clip will not restrict any movement or get in the way of either the lower or upper leaf during the actual shooting.

The STANLEY 902 hinge has about 3/4 inches of space between the leaves when the leaves are parallel to one another. This is important because you can close that distance some and begin shooting even before the leaves move into a parallel position! You usually can't do this with pin hinges attached to a wooden platform.

PIANO HINGES

One big advantage of a piano hinge is that it can be cut to the exact width of your platform. Essentially a continuous hinge, it will provide your tracker with stability across the entire upper leaf because it runs the entire width, rather than being spaced apart like pin hinges. You might be able to sight along the edge of this type of hinge for polar alignment. Most piano hinges have a narrow hinge line of about 3/16 inch and come in variable widths.

If you make a wide shooting platform or intend to shoot with telephotos in the 100-mm to 200-mm focal length range, consider this type of hinge as a first choice. You'll find it at most well stocked hardware stores.

PLASTIC HINGES

ASTRO KITS uses a polyolefin one-piece plastic hinge on its trackers. Besides being perhaps the lightest hinge material available, this plastic has such a thin hinge line (1 mm) that it can substitute for a finder. When you flex it back and forth, it seems as if it will break apart; but, for what it's worth, the manufacturer claims that its durable enough to take 69 million flex cycles.

This hinge (as well as many individual tracker parts) can be ordered separately from Astro Kits. (See Appendix for details)

BALL BEARING HINGES

Some refrigerator doors require heavy-duty ball bearing hinges. These large but smooth operating hinges could be used if you wanted a rock-solid, easy opening platform leaf for using medium to long telephotos. Each hinge weighs about 1 pound.

Contact National Manufacturing Co., Sterling, IL 61081, Phone (815) 625-1320 for prices. Currently, three ball bearing hinges are sold as a package and cost about $55.

CUSTOM MADE HINGES

If you're not satisfied with any of the commercially available hinges, customize one. After reading about Frank Zullo's and other's success with portable tracking platforms, Dennis di Cicco incorporated past tracking tips and ideas into a one-hinge metal platform machined to accept a small DC motor and a curved bolt. For the specifics on his personalized hinge design, read the Observer's Page in the October, 1985, SKY & TELESCOPE issue.

Regardless of the kind of hinge you buy, make sure that the upper and lower leaves move apart snugly but without binding. A loose hinge can only work against your star shooting efforts.

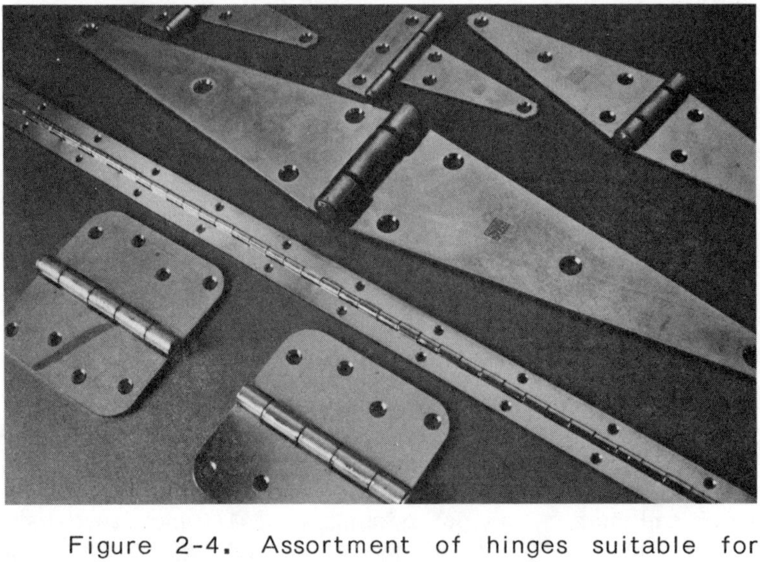

Figure 2-4. Assortment of hinges suitable for using on a star tracking platform.

THE DRIVING SCREW OR ROD

One- to three-foot long 6-32, 8-32 and 10-32 rods are available in the hobby-supply section of hardware stores. Select a few 10-32 rods that are arrow-straight. You'll probably be able to make at least two driving screws from each foot-long section, but get the extras for experiments. Don't buy the 6-32 rod -- it's too flexible and may not hold the weight of hefty camera systems. The 8-32 rod will work but won't be as strong as the 10-32 rod.

If you decide to use a 20 t.p.i. rod, there are numerous 1/4-by-20 bolts ranging upwards to about 6 inches. These have the advantage of being very strong and can handle more weight than a 10-32 rod. However, if you decide to curve the bolt (the subject of a later section), you may find it difficult to bend a 1/4-by-20 bolt as evenly as a smaller diameter rod.

After selecting a rod, spin a nut up and down its length to make sure the threads are cut cleanly and smoothly. Even the slightest imperfections in the threads could cause jerkiness in the mount's operation.

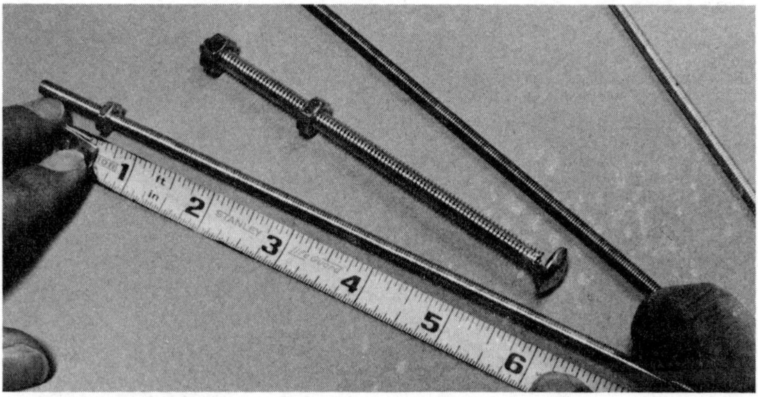

Figure 2-5. Use only the straightest part of the threaded rod for driving the tracker.

Buy a handful of washers, nuts, T-nuts, metal and nylon cap nuts. Also get a number of 1/4-inch, 1/2-inch and 3/4-inch long binder posts through which the rod can turn.

GEARS

Curved-bolt drives will require more hardware than straight-bolt drives. One of the most accurate ways of moving the upper platform is by using a small DC motor that drives a spur gear and main gear. The main gear should have at least 100 teeth for optimum performance; gears with fewer teeth won't mesh together tightly enough and will cause small but nonetheless detrimental movement.

For a gear to work properly, it must ride as friction-free as possible in a bushing. If you decide to build with gears, you must also purchase a bushing that you'll seat into the lower platform leaf.

The spur gear must mesh perfectly with the main gear and hold its position when tipped to the angle required for polar alignment. Be sure to test your gear-bushing-screw drive at the angle at which you will be using it. The teeth of the spur gear may ride up the main gear teeth when the shooting angle of the platform exceeds 50 degrees. You probably won't be taking your star tracker to a latitude like Iceland's 65 degrees, but test your tracker at a steep angle anyway. Better to overcompensate than to find out -- in the field -- your tracker's gears won't work!

Gears with a hundred or more teeth will have to be specially ordered. Even hobby shops that carry both nylon or plastic gears for model cars and planes won't stock this item. Astro Kits uses a 100-tooth custom made gear that you can order from them. Or, you can order one from a large company like Boston Gear, 14 Hayward Street, Quincy, MA 02171.

Another source for high quality gears and mechanical components is Winfred M. Berg, Inc., 499 Ocean Avenue, East Rockaway, N.Y. 11518. If you decide to build a gear-driven tracker, definitely order Berg's free catalogue of precision parts before starting your project.

The following photographs show two different gear systems in commercially available trackers. The illustrations may help you design your own gear drive.

Figure 2-6. The large worm gear in this Kajers mount makes one revolution every four minutes.

Figure 2-7. Both surfaces of these Astro Kits gears must remain parallel to ensure smooth movement.

PULLEYS AND WHEELS

The upper platform can be raised or lowered by using a pulley and belt drive. The pulleys and belt can be quite small. As long as they turn without any wobble or eccentric motion, you will achieve a smooth transfer of motion to the upper platform.

Replacement wheels for sliding screen doors make fine pulleys. Some of these come fitted with bearings that greatly reduce friction; others, without bearings, can be slipped onto appropriate ball bearings at minimum cost.

Prime Line Products Co., P.O. Box 2637, San Bernardino, CA 92406, makes screen door replacement wheels (No. B-551) that can be used as pulleys. These wheels (pulleys!) are 1 inch in diameter and have center hub openings that accept custom-made aluminum sleeves. These are actually binder posts, and they have an inside thread of 6-32 and are capped at one end. They come with the wheels. You get two wheels and two binder posts per package for less than $3.00.

The binder post caps can be drilled open and tapped to accept an 8-32 curved bolt or larger 10-32 curved-bolt. Since the binder posts extend all the way through the pulleys, you can easily seat the extended part into a bearing or work out your own system of attachment.

T-nuts are an excellent substitute for binder posts. Buy the kind that have a flush surface flange, not the ones with the points extending from the flange.

The T-nuts' centers are already tapped with common thread sizes and, once positioned in the wheel hole, can be screwed in place. Very strong, these inserts will withstand any pressure in the drive system and maintain alignment.

With pulleys you are adding a sideways force that's not present with a straight bolt system. You must balance and counter-balance the tension not only applied downward on the bolt, but perpendicular to it. These inherent problems can be worked out, but you will have to control these factors to make the drive work without error.

AN IMPORTANT POINT: If the pulleys wobble even slightly, longer lenses will often produce corkscrew images instead of sharp star images.

In a later chapter, you will learn how to "dry test" your tracker before turning it skyward. This will give you a chance to find out -- before driving 3 hours to a dark sky area -- if you will need to make any changes to the system.

BELTS

The belt could easily be the weakest link in a curved-bolt tracking system. Choose the correct belt, however, and you'll have no problems.

O-rings can work well with the pulley system mentioned above as long as you use an O-ring that is neither stretched too much or too little. Even though a number of O-rings might work on your system, usually only one will work flawlessly.

An O-ring stretched too tightly between pulleys will -- when the drive starts -- elongate AT ONE SECTION but not throughout the entire ring. It will stretch, say on the upper part of the pulleys, while the lower part is unaffected. If the pulleys are smooth running, the belt may even out in time; however, if there is uneven friction at some point, the belt will again stretch and pull in one place and not another.

This will drastically affect your astrophotography. The uneven stretching results in a kind of stop-and-go motion that will show up as star "blobs" -- ovals and spikes -- instead of points of light.

On the other hand, an O-ring fitting too loosely will, of course, just slip when the weight from the upper leaf is too heavy.

Once you've found the appropriate belt, buy a number of them and throw them in the bottom of your star tracker equipment bag. I once had an entire evening's shooting come to an abrupt halt when my drive belt broke and I had no replacement.

THE TIMING CYLINDER

During an exposure, you will be turning the driving screw at a precise speed, usually 1 r.p.m. To do this accurately, you will need to devise some kind of timer that is attached to the screw itself and that, when turned, gives the elapsed time.

The face of a kitchen timer works well because it's already marked at one-minute intervals. You can drill out the center and slip it directly onto the turning screw. If a kitchen timer face is used, the direction of the numbers usually dictates that you must lower the upper tracking platform rather than raise it. Or, simply paint over the face and reverse the numbers.

It's not necessary to have one second increments laid out on the timer. Two second markings may work even better because you will not have to strain to match, second by second, the real time to the "turning" time. One of my smaller star tracking platforms uses a wheel pulley surface as the timer; the time is shown in 5-second increments, with no loss in tracking accuracy when used with wide-angle or normal lenses.

I stumbled onto a perfect container to use as a timing cylinder. An OLD SPICE stick deodorant top has etched around its outer surface sixty evenly spaced marks. So easy.

A tip: smooth, shiny plastic should be your first choice material because it eliminates having to throw a shadow onto the turning surface for marking the passage of time. The beam from a cellophane-covered flashlight creates a pinpoint reflection where it strikes the plastic surface. You can easily watch the numbers pass through this highlight to mark the elapsed time.

TURNING HANDLES

A number of astrophotographers using star trackers prefer to hand turn the driving screw. They feel that they have more control, second by second, than they would with a motorized platform. They know that at the end of each minute's exposure, they will be perfectly synchronized with the timer; they can't be absolutely sure that will happen with a motor.

In order to match the master clock with your marked timer, you will need to choose a type of turning handle that is smooth operating, and works with a minimum of tension as it either lifts or lowers the upper platform. Many star trackers simply use a wooden paint stirrer that has been notched at each end for better finger control.

You can add to the stirrer a revolving handle -- like one you'd find on a fishing reel -- if you like holding onto the turning device rather than moving it with one finger.

It's extremely important for you to find out the way you feel most comfortable turning the driving screw. On long exposures, say beyond 12 minutes, your attention may drift and it will be up to your hand to take over automatically and continue through with the correct timing. When you have used your tracker a great deal, you will be able to move unerringly through these long exposures with little deviation from the required time.

LIGHTS

You will need some type of light in order to read the elapsed time during exposures. Buy a small flexible neck lamp that runs on two AA batteries. The flexible neck allows you to not only see the time on both the master clock and the tracker's timer, but it will let you quickly see your camera settings. Fit the lamp head with red plastic or cellophane to let your eyes remain adjusted for dark night conditions.

The flexible neck lamp can be removed quickly from the tracker when you need to rummage through the camera bag for equipment. It can just as quickly be readjusted on the moveable hinge leaf for timing purposes. Buy one at an auto supply store.

An LED, a light emitting diode, can be screwed in place on a tracker and work as a timer light. Some LEDs are less than 1/4-inch long and are feather light. These miniature lights usually draw little current and will run many hours before dimming.

Helmet lights and lights you can clip onto clothes are excellent because they free your hands. A helmet lamp is particularly useful when you have to walk into your shooting site carrying an armload of equipment.

Get an illuminated stopwatch. Next to the flexible neck lamp, the stopwatch light will get the most use.

Radio Shack makes such a wristwatch. It's called the Micronta Digital LCD Chrono-Alarm Watch and is often discounted below $10. This easy-to-set, easy-to-read stopwatch -- with timing ranging from hours to hundredths of a second -- eliminates fumbling about with one hand on the watch and one on a flashlight.

Always carry fresh batteries (check to make sure they are fully charged) and extra flashlights. Don't let your sole light source and batteries die with no backup available.

PHOTOGRAPHIC ACCESSORIES

BALL-AND-SOCKET TRIPOD HEADS

A ball-and-socket tripod head sits atop the upper platform of a hinge type mount and supports the camera and lens. Here are a few specific models you may want to consider.

1) The Leitz large ball-and-socket. Pricey (the current price is in the neighborhood of $110) but strong and smooth operating, this tripod head might be your first choice if cost is not important. One advantage of the large Leitz head is its height.

It measures 5 inches from its base to the camera mount screw on top, a distance large enough to allow the camera to be flipped far over on its back to allow a tremendous angle of view. This length also permits it to be rotated without the back of the camera bumping into the hinge or possibly the finderscope. It will rigidly hold up to 200-mm focal length lenses.

2) The Groschupp medium ball-and-socket. This model is much less pricey (about $25 currently) but is also rugged and smooth working. It is shorter,

measuring 4 1/4 inches from base to screw, but is a few ounces heavier than the Leitz.

The main post has an annoying characteristic of being free-spinning whenever the ball head is loosened. This makes it difficult to make tiny adjustments. Nevertheless, it is an excellent value at that price, and should be seriously considered. This tripod head will also hold lenses up to 200 mm in focal length.

3) SLIK makes a series of ball heads that will perform well on a tracker. Consider getting their Compact Ball Head (Catalogue no. 618-710) that costs about $28. Although shorter than either the Leitz or Groschupp, it will hold lighter cameras and lenses adequately.

4) If you are going to be using light-weight lenses, KALT distributes a German made medium ball head (Catalogue no. 035642) that is less expensive than the other three but nonetheless shows fine workmanship. This model presently retails for about $11.

There are some ball-and-socket heads made completely -- ball and all -- from a tough plastic-like material. Don't buy one of them. The ball slips too easily, even on light camera equipment. You are better off buying one of the ball heads mentioned above.

Ball heads are the best choice for use on the upper platform. However, you should use a standard pan-tilt tripod head for holding the entire star tracker platform.

THE BASE TRIPOD AND PAN-TILT HEAD

The tripod and head supporting the tracker must be rigid and strong. Here are a couple of recommendations.

Gitzo manufacturers over a hundred different kinds of tripods which are professionally machined, easy to use, and rugged. A tripod in the 200 series, specifically models 220 or 224, would be an excellent choice. The legs alone will cost about $150.

Consider buying the Gitzo legs only, and add a Bogen 3029 pan-tilt head (about $30) for a system that will firmly hold any tracker you might make.

Tiltall is another excellent tripod used by many professionals. It comes complete with pan-tilt head and costs about $125.

Some astrophotographers like Frank Zullo mount the tracker, or wedge beneath the tracker (see Figure 3-4), directly to the tripod legs for even greater rigidity.

I use a Gitzo 220 tripod and mount my trackers directly onto a Gitzo short centerpost. I keep this 6-inch centerpost continually attached to a tracker and can quickly slide it into position when needed. A short centerpost for a Gitzo 220 tripod costs about $25.

MISCELLANEOUS TRACKER ACCESSORIES

BATTERIES

REPLACE IN-CAMERA BATTERIES WITH EXTERNAL ONES

Long time exposures required for astrophotography will drain camera batteries quickly. You are fortunate if your camera has a mechanically activated shutter that will work independently of battery power. Just remove the batteries -- you certainly won't need them for taking meter readings.

If your shutter requires batteries, you can replace the internal camera battery (or batteries) with an external series of batteries. Fit a battery holder on the underside of the tracker platform and run wires from it to a "dummy" battery in the camera battery chamber. Insert four AA alkaline batteries (or the correct number for your camera) in the holder. This setup will let your shutter work much longer before the power gives out.

A LIGHTWEIGHT 6-VOLT SOLUTION

Polaroid makes a wafer-thin, extremely light 6-volt battery that can run an automated tracker for hours. Dennis di Cicco uses one on his single-hinge platform and says it performs well.

There are two types, the Standard (P100) and Lithium (P500). The Standard costs $3.50 plus $1 shipping; the Lithium is $5.25 plus $l shipping. Order a few of the Lithium (for longer life) if you need to travel lightly and intend on shooting in moderate temperatures. Available from: Power Card Corporation, Brookline Street, Newton, MA 02159.

A 12-VOLT SOLUTION

Tower Hobbies, catering to enthusiasts of radio-controlled cars, boats and airplanes, offers a nifty catalogue called Tower Talk that's filled with kits and tools and parts needed to get these "super toys" together and running.

One of its best bargains is a maintenance-free, rechargeable, sealed, lead-calcium 12-volt battery that delivers a full 6 amp-hours of power. It measures 5.9" x 2.5" x 3.6" and weighs 5 1/4 lbs. Because it's sealed, there is no chance of chemical spills or explosive gas leakage. It's called the MAFREC. You should definitely consider getting one. I use one on my trackers and it performs beautifully and is simple to recharge.

Tower also sells a 2/6/12-volt charger that charges the MAFREC overnight. It comes with a 6-foot cord with alligator clips, resistors, and is U.L. approved.

The MAFREC battery (Stock Number MAFP1000) is priced at $27.95. The 2/6/12-volt charger (Stock Number TOWP1000) costs $8.95. This combination -- and price -- is hard to beat.

Send for their catalogue ($3 or free with an order) that's packed with goodies like ball bearings, wheel collars, ball links and other parts that might prove indispensible for your tracker (or some other) project. The catàlogue's a "wish" book and a great way to buy hard-to-find items.

Their address: Tower Hobbies, P.O. Box 778, Champaign, IL 61820. Toll free order line: 800-637-4989.

Celestron, Orion, Kajers, Roger Tuthill and other telescope and telescope accessory outlets offer 12-volt rechargeable batteries. With any commercial tracker that runs off 115 V-AC, an AC-DC converter may be required for field use. Call one of these companies for specifics.

DRIVE CORRECTORS

The trackers described in this book are intended to be used for unguided astrophotography. Some commercially available mounts come with guide scopes that let you make minute adjustments during the exposure.

If your tracker can be guided, a drive corrector -- or something similar -- will be required to keep the guide star centered on the reticle of the optics. Kajers, Orion, Celestron, and a number of other companies make these units. When you think about spending this kind of money ($150 to $400) to help you track better, you might be better off buying a telescope with a good drive system. Just piggyback the camera on the telescope.

MOTORS

Motors are covered in the section dealing with automating a drive. Specific models, along with prices and addresses of distributors, are presented there.

Figure 3-1. You can make this straight-bolt wooden tracker by following the plans in the next few pages. The wedges shown here are optional, and are discussed at the end of the chapter. An automated version of this mount is shown in the GALLERY section.

ASSEMBLY

In the last chapter, you were offered a shopping list of parts that can be used on trackers. Now the joy of building begins.

I've presented a basic design for a straight-bolt wooden tracker, a straight-bolt metal tracker and a curved-bolt metal tracker. At the end of each of these projects I've covered in some detail modifications that you may want to make. The basic models will perform well; the refinements might help them work even better.

These are hinge-type mounts that will track at the sidereal rate. Each of them may be automated, a subject covered later.

I've listed the distributor and parts number of items essential for each project. In many cases, you can substitute similar parts for the ones mentioned.

Using a properly assembled straight-bolt wooden tracker, you can expect to take 10- to 15-minute (and even longer) non-trailing star shots with telephotos in the 100-mm range. With precise polar alignment, you can use 200-mm or 300-mm lenses for 6 to 8 minutes before star trailing sets in.

There is a certain beauty in using simple designs to capture the grandeur of the night sky. These trackers will give you many hours of contentment, from the assembly stage to using the finished product.

MAKING A STRAIGHT-BOLT WOODEN TRACKER

You'll be able to make this tracker in a few hours with simple hand tools. The basic design is quite similar to G. Y. Haig's original mount.

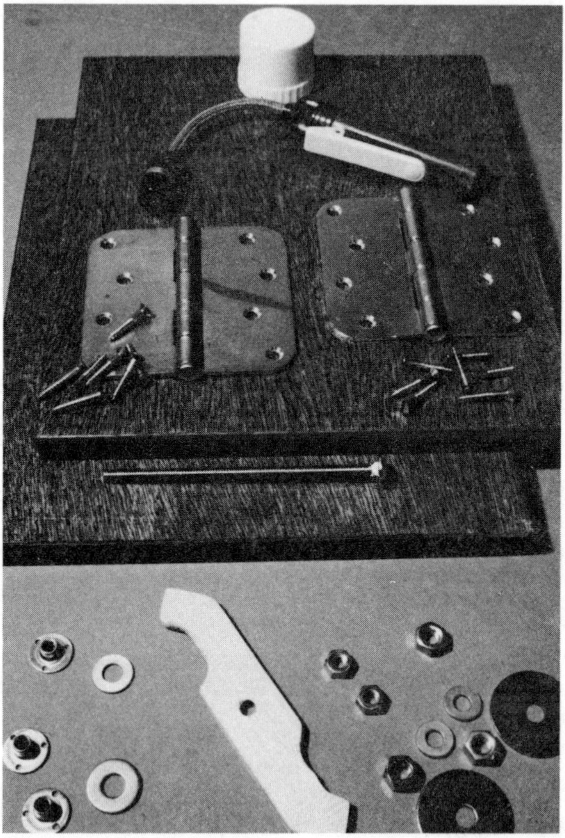

Figure 3-2. Parts to make a wooden straight-bolt tracker.

This model uses a 1/4-by-20 straight bolt to lift the upper leaf. Other size bolts may be used, but you will then have to re-calculate the correct distance from the hinge axis to the bearing point of the rod.

PARTS LIST

2 Pieces .... 10-by-12-inch plywood, any grade, three-quarter inch thick
2 .... 3-inch door hinges
1 .... 6-inch, 1/4-by-20 carriage bolt
2 .... T-nuts, 1/4-by-20 inside diameter
1 .... Turning handle: your own choice
1 .... Timing cylinder: your own choice
1 .... Ball-and-socket head: your own choice

Miscellaneous: a stick-on dashboard digital clock, flexible neck clip-on lamp, handful of fender washers, lock washers, wood screws to mount the hinges, a polar alignment finder and mounting of your choice.

ASSEMBLY

Attach the two hinges along the 10-inch sides of the boards so that they open evenly, tightly, and smoothly. On the bottom board, measure exactly 11 7/16 inches from the hinge axis. This measurement has to be precise in order to lift the upper board at the sidereal rate.

Drill a hole at this point, countersinking one of the T-nuts. The 1/4-by-20 carriage bolt will turn through this T-nut.

Position the carriage bolt so that its rounded head presses against the lower surface of the upper platform; the other end of the bolt, after passing through the T-nut, can be adorned with any type of turning handle: paint stirrer, custom-made, notched out handle, or any other aid to help turn the bolt.

To mount the ball-and-socket head atop the upper platform, run a 1/4-by-20 bolt through the wood and into the base of the ball head. Exact positioning of this tripod head is unimportant; in fact, you may want to add extra attachment points to be used with different camera/lens combinations to get the best positioning over the base.

Secure the second T-nut to the bottom of the lower platform. Attach a sturdy camera tripod here.

Use your inventiveness to install the small light, the timing cylinder and the digital watch to the underside of the mount.

The timing marks will get smaller from left to right: ... 60 . 58 . 56 ... if the bolt is to LIFT the upper leaf.

Figure 3-3. Details of various timing cylinders and turning bolts. Much of the success of this type of tracker lies in its simplicity.

MODIFICATIONS AND REFINEMENTS

Frank Zullo's own wooden timing cylinder has marks spaced in a way that allows him to speed the drive a bit as the exposure time increases. This is to compensate for errors naturally occurring when operating a straight-bolt drive during exposures exceeding 20 minutes. But, for most wide angle or normal shots with exposures of 10 minutes or less, no timing compensation is necessary. If the platform is accurately polar aligned and the tracking is smooth, no star trailing will be visible with such lenses.

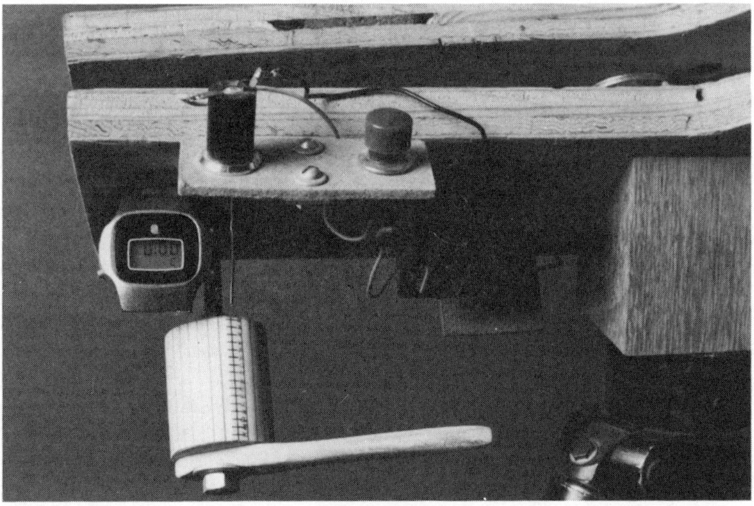

Figure 3-4. A straight wire casts a shadow onto the timing cylinder. Thin horizontal lines mark the passing of minutes. Also note the wedge support above the tripod. Tracker design and photo by Frank Zullo.

Check where the top of the carriage bolt presses against the upper platform. (For more details, see the mechanical testing section) If the carriage bolt wobbles as it lifts, you may need to replace the top with a 1/4-inch cap nut. Just cut off the carriage bolt head and screw on the cap nut.

If the cap nut wobbles during lifting, remove it and file the bolt end to a small, rounded, centered point. Have it bear against a washer on the underside of the upper platform. This should eliminate any "corkscrew star trail" problems in this straight-bolt system.

To align this tracker, you should add some kind of finder device ON THE UPPER HINGE LEAF. If there is any board flexure caused by camera/lens weight, the finder will at least move with this change.

To track, turn the handle at exactly 1 r.p.m. matching the clock second by second. The ball-and-socket head permits you to aim and lock the camera at almost any skyward position.

Mr. Zullo has experimented with variations of this basic design and has made a few personal modifications:

"I have taken off the light that was used to illuminate the sighting device. It is just as easy to use a red-filtered flashlight, and eliminating the light has given me one less thing to worry about.

"With my first mounts," he said, "I was achieving consistent tracking on only my 10-minute exposures with a 100-mm lens. Longer, non-trailed exposures were very inconsistent.

"After experimentation I found it was due to movement in the tripod head -- flexure. I removed the head and centerpost of the tripod and bolted the tracker directly to the flat base of the tripod. In doing this I lost the tilt/pan controls, but got around this by putting a 4-by-4 block between the tripod and tracker that was cut to an angle equal to my latitude -- 33 degrees. (See Figure 3-4)

"The pan/tilt movements for alignment come from raising or lowering one of the tripod legs (one pointing directly south) and gently moving the other two tripod legs to either side (hence, actually swinging the tripod around). These movements may seem crude, but they are actually more precise and easier to make than pan/tilt movements with the old tripod head."

The wedge type of support that Mr. Zullo mentions above was, of course, used with some earlier star mounts but was abandoned in favor of lighter, more portable tripod systems. But, I agree with Mr. Zullo's point, and have found from experimenting that a wedge, mounted directly atop the tripod legs, increases rigidity. Another wedge can be screwed to the top plate to keep the ball-and-socket tripod and camera/lens over the main tripod's center of gravity. This design puts less strain on the hinges. (See Figure 3-1)

You might make the base wedge extra large to add weight. Not as portable as the basic design, the added support should permit moderate 100-mm or longer telephotos to be used for short exposures.

"The overall gain in rigidity from using the wedge now allows for consistent 20-minute exposures with a 100-mm lens," Zullo said. "In a test I did once, I got a perfectly tracked 25-minute exposure with the 100-mm lens. Usually I don't like to push it, so I have settled on 20 minutes.

"I built this tracker with a turning bolt of 30 minutes tracking capacity. Hence, I have gone to 30 minutes with wider lenses, usually the 50-mm."

The longest lenses that Mr. Zullo has used successfully on a wooden straight-bolt tracker?

"I have used a 200-mm lens for up to 7-8 minutes, and a 300-mm lens in special applications. When using these longer lenses, I think it would be possible to gain longer exposure times. I feel a lower geared drive would make things a lot easier, since this would allow for smoother tracking because movements would be in smaller increments. This is where the constant speed of a motor would help greatly."

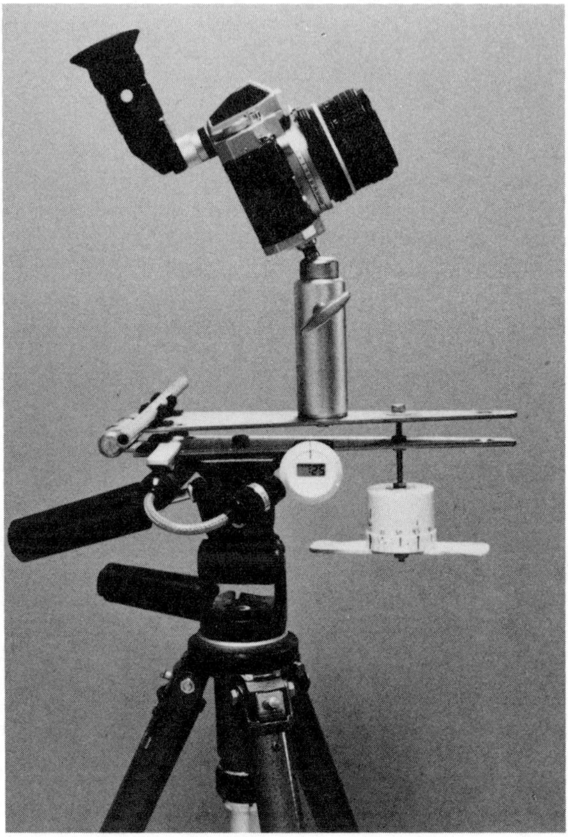

Figure 4-1. A straight-bolt metal tracker that uses a plastic water-supply tube for polar alignment.

MAKING A STRAIGHT-BOLT METAL TRACKER

In this design, one hinge is the entire tracking platform. As mentioned before, choose a tight but smooth-working 10-inch Stanley (R) V902 strap hinge with NYLON SPACERS around the hinge knuckle. If you need to take up play in the hinge, tap a 5/8-inch "E" clip next to one of the spacers.

THE PARTS LIST

1 ..... 10-inch Stanley (R) V902 or 902 strap hinge
1 ..... 12-inch 10-32 t.p.i. straight rod
2 ..... 1/2-inch long binder post with 10-32 threads
1 ..... Box of Velcro or Velcro type fasteners
1 ..... 12-inch gray plastic vanity water-supply tube
1 ..... Wooden paint stirrer
1 ..... Plastic or wooden cylinder for timer marks
1 ..... Ball-and-socket tripod: any type

Miscellaneous lock washers and nuts, a handful of different length 1/4-by-20 bolts, an auto dashboard clock, a flexible-neck clip-on lamp. A finderscope is optional.

The 10-32 turning bolt was chosen because of the length of the hinge. A 1/4-by-20 bolt CANNOT be used in this project to get a 1 r.p.m. sidereal rate.

Figure 4-2. Parts needed for building a straight-bolt metal tracker.

ASSEMBLY

Please note: electric hand drill bits can bind when cutting through metal, so play it safe and clamp the hinge to a piece of wood before drilling. Better yet, use a drill press.

On the lower hinge leaf, centered from edge to edge, drill a 1/4-inch hole about 3 inches from the knuckle. The main tripod will attach here.

Now -- also on the lower leaf -- accurately measure out from the center of the knuckle 7 9/64-inches and drill a hole a bit smaller than the outside diameter of the binder post (Figure 4-3). (IMPORTANT NOTE: If the hole is drilled closer in or farther away from the knuckle, the tracker will not be accurate.) File the hole so that the binder post fits into it snugly.

Cement or Super Glue the binder post, flange side up, into the hole.

Figure 4-3.

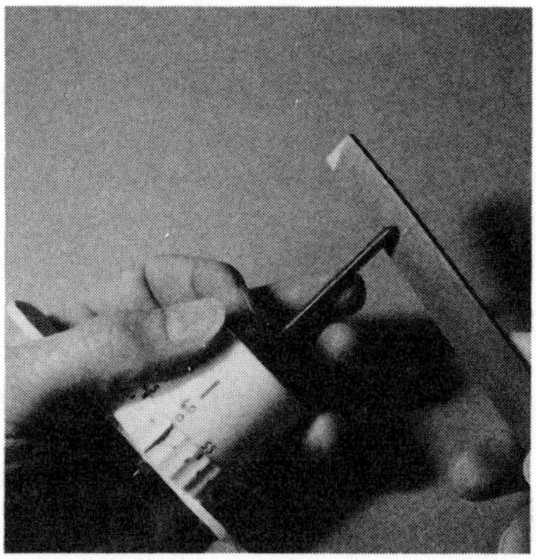

Figure 4-4.

Cut about 7 inches from the 32 threads-per-inch rod and file one end of this section to a rounded point (Figure 4-4). (IMPORTANT NOTE: make sure that the rounded tip is EXACTLY in the center of the rod. Put the other end of the rod in a drill chuck and spin the rod to check this out. If the point wobbles, even a tiny amount, the upper leaf will not lift smoothly and you will get corkscrews instead of pinpoint star images.)

Be sure the rod is long enough to go completely through the timing cylinder and the paint stirrer -- and still have an additional 3 inches to the point.

Divide the timing cylinder into 60 equal parts around the perimeter. Like the timer for the straight-bolt wooden tracker, the numbers should get larger as they progress to the left if the bolt is to lift the upper platform.

Drill a hole through the center of the timing cylinder large enough to accept the rod. Twist the rod through the hole, pointed end up, leaving about 3 inches of the screw above the cylinder.

Fashion the paint stirrer into a turning arm 4 or 5 inches long. Notch each end for better control. You can attach the handle of an old fishing reel to the turning arm for even smoother tracking.

Drill a hole through the center of the turning arm and fasten the arm onto the rod below the cylinder.

Stick Velcro strips to the top of the hinge leaf, about an inch from the knuckle. Place Velcro on the water-supply tube. Now, position the edge of the tube parallel to the center of the hinge. The tube can now be used for polar alignment.

Tape red cellophane over the end of the flexible-neck light and clip the light to the bottom leaf, next to the hinge knuckle.

Stick another strip of Velcro onto the main tripod just under the platform. Place the small auto dashboard clock onto this strip.

Attach the ball-and-socket tripod head to the top leaf with a 1/4-by-20 bolt that will screw into the tripod's base. (The Stanley Company has conveniently provided a center hole for the attachment!)

Rejoice! You've just made a straight-bolt tracker that is portable, smooth operating, and easy to set up. Shooting under similar conditions -- say, unguided shots for 15 minutes using a 100-mm lens -- your tracker will run neck and neck with most commercial trackers.

MODIFICATIONS AND REFINEMENTS

With repeated use, the tip of the rod may twist into the underside of the top leaf and drill out a path. One solution to this problem is to let the rod tip bear against a washer on the underside of the top leaf.

Just turn the washer (held by a 1/4-by-20 short bolt and nut) a 1/16 inch or so after each "run" of the bolt across the washer. You will give the rod tip a new surface for every lift of the upper hinge.

Another modification is to drill a hole into the UPPER plate where the turning rod should bear. Make the hole 1/4-inch larger than the rod's diameter.

Cement an axle nut to the top surface directly over the hole. The turning bolt's tip goes through the upper plate and bears against the axle nut. The edge of the hole will keep the rod precisely in place when heavier telephotos are used.

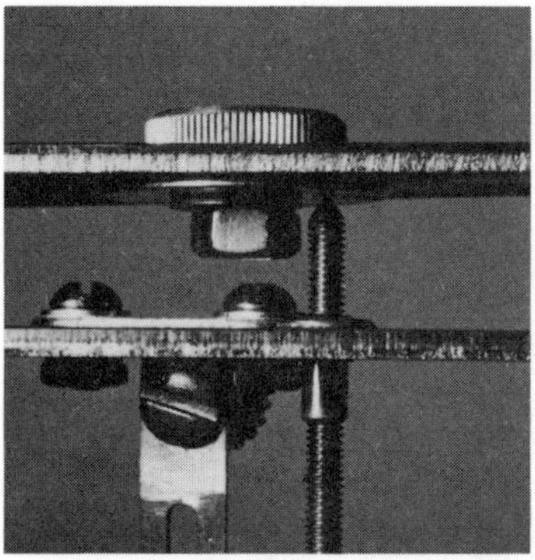

Figure 4-5.

Figure 4-6.

Figure 4-5. Rod tip turning on a washer rather than the hinge surface. Figure 4-6. Axle nut holding the bearing point of the rod.

Figure 5-1. This curved-bolt metal tracker eliminates tangent error and is quite portable.

MAKING A CURVED-BOLT TRACKER

"It won't work," my astrophotographer friend said when he saw this design. "It's too small and will lack the precision needed for taking good astrophotos."

He changed his mind after seeing a few shots I'd taken with it. Hundreds of stars in the Milky Way stood out crisp and tack-sharp on the slides. He just smiled and said, "Well, it didn't look like it would work."

You can make this tracker using easy-to-find, off-the-shelf parts. It's an extremely small, lightweight, hand-driven model. Using an 8-inch Stanley strap hinge (as the entire platform) you will be able to employ a belt-and-pulley, curved-bolt design for wide-angle and normal lens use.

The tracker's advantages: weighing only 12 ounces (less the ball-and-socket head and light source) and measuring 3 inches by 8 inches, this mount can be tucked into the smallest camera bag. Because it's hand driven, there are no batteries, wires or other electrical components to carry about. The platform will track stars accurately with wide angle and normal lenses for at least 10 minutes. (A larger version of this tracker is in the AUTOMATING A CURVED-BOLT DRIVE section.)

Figure 5-2. Parts needed for a curved-bolt metal tracker can easily be adapted to wooden mounts.

The drive's disadvantages: the upper leaf's small surface area lacks the structural support and stability of bigger trackers. Lenses over 50 mm will track only for short times (up to 4 minutes with a 105-mm lens) before tracking problems occur and the stars begin to lose their roundness.

Turning a crank below the bottom leaf turns a pulley above. A belt from this pulley turns a second pulley that has a nut embedded in its center. The turning nut rotates up a curved bolt which is attached to the top leaf and freely slides through an opening in the lower leaf. The rotating nut lowers the upper leaf at the sidereal rate, giving you exposure control.

THE PARTS LIST

1 ..... 8-inch Stanley strap hinge (902)
1 ..... 12-inch 10-32 t.p.i. rod
1 ..... Package of screen door rollers (B-551)
1 ..... Wooden paint stirrer
1 ..... 12-inch vanity water-supply tube
1 ..... 1 ball bearing with 1/4-inch I.D.
1 ..... O-ring (depends on pulley design)
Miscellaneous washers, 1/4-by-20 screws, nuts, an auto dashboard clock, a flexible-neck clip-on lamp

The Stanley 8-inch hinge has a predrilled hole about 7 3/4 inches from the hinge knuckle. File a larger opening whose center will be at 7 9/64 inches. You will eventually screw down one end of the curved bolt on the top leaf, and position the other end so that it will slide smoothly through the lower opening.

The screen door rollers (B-551; Prime Line Products Company, P.O. Box 2637, San Bernardino, CA 92406) come with binder posts. Insert one into the center of a roller; the end will emerge about 1/4 inch. This extension should fit snugly into the center of the ball bearing. File or sand the post until it does.

Cement the post to both the inside of the pulley wheel and the inside of the ball bearing. Drill a 1/4-inch hole through the bottom leaf 4 3/4 inches from the center of the hinge knuckle. The bearing and pulley will sit here, centered atop this hole.

Here's a challenging part -- finding a way to secure the bearing in place while still allowing it to rotate smoothly. My job was fairly easy because the outer rim of the bearing I was using was a tiny bit thicker than the inner, moving section. After centering the pulley above the hole, I simply cemented the outer lip to the metal surface. It is strong enough to counter the belt tension and still allow the pulley to move effortlessly. If your bearing is the same thickness throughout, you may have to add a bushing on its outer edge.

To finish this pulley assembly, insert a 1-inch to 2-inch section of 8-32 rod through a paint stirrer handle and into the bearing shaft. My handle is about 2 1/2 inches long and uses a fishing reel for easier turning. Be sure the handle won't bump into the curved driving rod.

Bend the 12-inch section of 10-32 t.p.i. rod into the required curve. Use a pattern made by swinging a circle whose radius is 7 9/64 inches. Keep laying the curved rod on top of the pattern to gauge your bending progress. (A tip: to get an even bend, press the center of the rod against something like a street lamp post.)

Saw 5 inches of the best portion of the curve from the rod to use on the tracker. This will give you plenty of star shooting time.

Measure the thickness of the ball bearing. You'll need to use a spacer the same thickness as the ball bearing in order to keep the belt (O-ring) parallel to the bottom leaf and riding correctly in the pulley grooves. A scrap piece of Teflon should work well, providing a slick surface for contact. Or, try plastic chair-leg glides. Another solution would be a bushing sunk into the metal leaf.

The center of the spacer must be drilled out so that when the rod lifts and lowers, no contact takes place inside.

Insert a binder post in one of the pulleys. Have the binder post flush with the top and bottom of the pulley, so cut off the capped head accordingly.

Cement or Super Glue the post to the inside of the pulley. Do this as you turn the pulley up and down the rod to make sure it spins as smoothly as possible. If there is any wobble at all, throw the pulley out and use another. Even the slightest off-center movement will affect the drive enough to cause star blurring or corkscrewing.

Choose an O-ring snug enough to maintain pressure on each pulley without slipping. Too tight, and the O-ring will exert too much side force; too loose and the O-ring will either not grab the wheel or periodically slip. The one I use is a McCord water outlet gasket, part number R4694, 74-0096. You'll find it at large auto supply stores.

You can make a small, swing-out platform (actually a 1-inch long paint stirrer section with Velcro taped to it) that will hold your watch timer. It can be swung under the platform when not being used.

Place Velcro strips on the top hinge, parallel to the hinge axis. Put Velcro around your finder -- the 12 inch vanity water-supply tube.

Drill a 1/4-inch hole in the lower hinge leaf about 1 3/4 inches from the center of the hinge knuckle. The tracker will mount to the tripod here.

Now for the timing device. Divide the top side of the outermost pulley into quarter units, marking the units "0", "15", "30", "45" in a clockwise direction. Evenly space two dots between these units. Make a timing mark on the hinge itself.

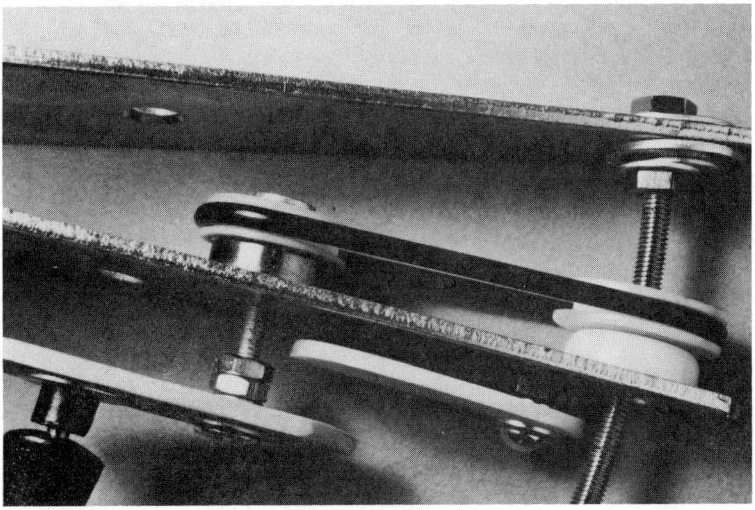

Figure 5-3. Details of pulleys and turning handle on a curved-bolt metal tracker.

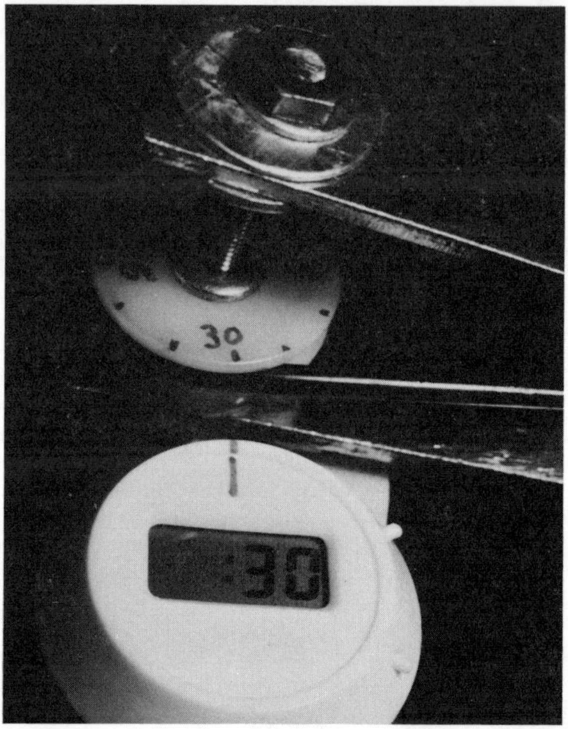

Figure 5-4. Close-up of timing marks on pulley and on watch face.

ASSEMBLING THE PARTS

Spin a nut and lock washer about 1 inch onto one end of the curved bolt. Insert that end of the bolt through the end hole of the upper leaf. Spin a lock washer and nut onto this end.

Adjust the curved bolt until its other end moves cleanly through the center of the hole in the lower spacer and hinge. Make sure that the rod never hits anything as you open and close the hinge.

You may have to bend the upper tip of the rod minutely in a direction exactly opposite of the curve in order to tighten down the nuts. Do this only if the rod keeps bending too much toward the hinge axis.

Now look directly down at the hinge and make certain that the rod, as it moves through the hole, is 90 degrees from the hinge axis through its entire length. In other words, the rod should be precisely adjusted in two different planes: the pulley on the rod must sit flush with the lower spacer, and the rod must move perpendicular to the hinge axis. It sounds harder than it really is. As long as everything looks right to you from every angle, you've got it right. Tighten the nuts in place when everything is perfect.

Stretch the O-ring around both pulleys. Insert the handle and give the tracker a few turns. The action should be without friction and, if everything is positioned correctly, the upper platform should lower effortlessly.

Mount the timer on the swing-out stand and tighten down a ball-and-socket head on the top leaf. Clip on the light. You're ready to go.

MODIFICATIONS AND REFINEMENTS

Whenever a heavy lens replaces a lighter one on the tracker, the extra weight means more friction. More friction usually means a change in speed if the drive is automated, or at least more resistance through the turning handle if the drive is hand turned.

One way of greatly reducing the added friction is by using a spring. Ajax Wire Specialty Co., Inc., 119 Bloomingdale Rd., Hicksville, N Y 11801, makes an 11-inch long, 1/2-inch diameter spring (Part no. 62) that has just the right amount of flexibility to use on trackers. But, you don't use it in the usual fashion.

On one of my curved-bolt drives using a 10-inch Stanley hinge, I connected the ends of the spring to the ends of the upper and lower leaves of the hinge. This, of course, makes the spring bow outward, and spreads the leaves apart. By simply turning the spring backwards toward the hinge knuckle, I can find the precise position where any combination of camera and lens allows the pulley to just touch the bushing. It doesn't matter whether I'm using a light 24-mm lens or a hefty 80 - 200-mm zoom lens; I can adjust the spring until the pulley driving the upper leaf glides across the lower surface. I don't even have to adjust the motor's output after I adjust the spring.

The spring has another advantage. It keeps the upper leaf from inadvertently flying open -- possibly throwing your lens against the tripod -- while you are walking with or setting up a tracker. This can be a problem with either a straight-bolt or curved-bolt system.

In strong wind, the spring may sometimes vibrate, causing havoc with your shooting. That's the only disadvantage I have found using the spring. Just remove it on those blustery nights.

Try the spring or a pair of springs on a larger, wooden platform. Springs attached at the corners of the upper and lower leaves may even make the tracker a bit more rigid.

Another way of reducing friction between the pulley and lower leaf bushing is by using a thin thrust bearing. A thrust bearing looks a bit like a fender washer, but has tiny rollers spread evenly around its surface. Every roller extends slightly above and below the bearing. Any surface contacting this bearing will ride upon these rollers and move smoothly over the other surface. A perfect, inexpensive, lightweight solution to a friction problem. The bearings I use have an outside diameter of 1 inch and an inside diameter of about 5/8 inches, but many other sizes are available at gear and bearing companies. The Torrington Company, 59 Field Street, Torrington, CT 06790 is an excellent source for specialized thrust bearings.

MOTORS AND MOTOR CONTROL UNITS

AUTOMATING A STRAIGHT-BOLT DRIVE

In the previous section, we covered the materials needed and the step-by-step procedure for making hand-driven trackers. This section will show how you can easily modify those designs and make automated platforms.

A correctly mounted motor will reduce vibration during shooting, making longer exposures possible.

TYPES OF MOTORS

A DC gearmotor or stepping motor can be used on a star tracking mount. A gearmotor can have a high torque rating, be lightweight and be powered by small, portable batteries.

A stepping motor usually has great rotational accuracy and, like a gearmotor, is portable, powerful, and dependable. Because this type of motor is currently being used in data processing and robotics, it offers star trackers an inexpensive and reliable means of automating a platform.

The Thomas Register, a reference book series found in many libraries, lists scores of distributors of small DC stepping and gearmotors -- as well as synchronous motors -- suitable for automating drives. Use this excellent source to find a motor perfect for your needs. Call the distributor for specifics.

A motor listed in the Thomas Register and already being used by many star trackers is a high torque gearmotor made by the Hankscraft Company and distributed by Edmund Scientific, 101 E. Gloucester Pike, Barrington, N.J. 08007. This motor has a stall/torque rating of 65 inch-ounce for its 1 r.p.m. model. You can specify whether you want the motor to run off either a 3-volt or 12-volt DC battery. Check Edmund's catalogue for specifics.

USING THE HANKSCRAFT MOTOR WITH NO MOTOR CONTROL

If you plan to use a Hankscraft motor (Part no. F41,860 for a 12-volt 1 r.p.m. model; Part no. F41,327 for a 3-volt 1 r.p.m. motor) without modifying its turning speed, you must know the exact r.p.m. rate, not what's stated on the motor.

If the load atop the tracker won't vary too much, AND you will be shooting in moderate temperatures, AND the battery power will always remain constant, then heed the advice of Frank W. Loflin II. In the October 1984 issue of SKY & TELESCOPE, Mr. Loflin writes that in order to know where to drill the hole for the driving bolt, you must first multiply the motor's actual r.p.m. rate by 228.6 before dividing by the bolt's threads per inch.

I'd like to modify Mr. Loflin's point and say that you should determine the actual r.p.m. rate while the motor is under load, lifting a typical camera/lens combination. This will give a more accurate r.p.m. rate than when measuring the motor under no load.

Let's say your motor shaft turns once every 61.4 seconds under load. 60/61.4 x 228.6 = 223.4. If the turning bolt has 32 threads per inch, the nut must be placed 6.9 inches from the hinge axis.

The Hankscraft motor is so light you can attach its shaft directly to the end of the tracker's driving bolt and, with a guide, let the motor push the rod's point against the upper platform. But before you make this modification, build some kind of motor control to get consistent timing for longer astrophotography.

Here are some options:

A SIMPLE VOLTAGE REGULATOR

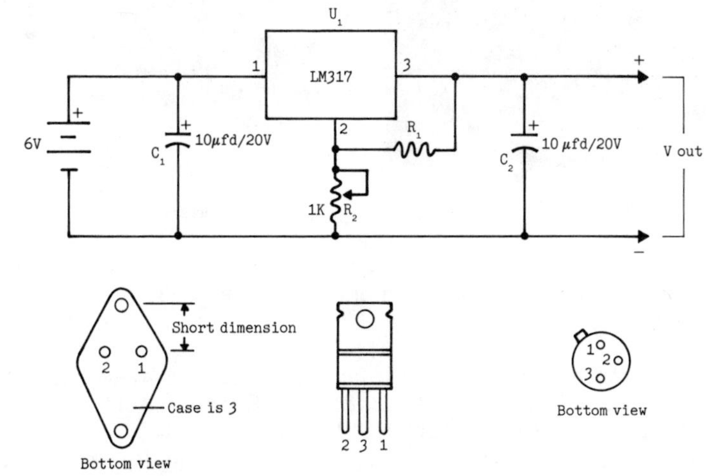

R1	390 Ω 5% ¼ w Resistor (carbon film)
R2	1K Potentiometer
U1	LM317 Adjustable voltage regulator
C1, C2	10 μfd Aluminum electrolytic cap 15 WVDC

$$R_1 = \frac{1.25\ (500)}{1.7} = 368\ \Omega$$

$$V_{out} = 1.25\left(1 + \frac{R_2}{R_1}\right) = 10^{-4}\ R_2$$

Min output ≈ 1.25 V

For R_2 at midrange (500 Ω), $V_{out} = 3V$

$$\text{Max. output} \approx 1.25\left(1 + \frac{1000}{368}\right)$$

$$3 - .05 = 1.25 + \frac{1.25\ (500)}{R_1}$$

≈4.6 V (If input voltage greater than about 6.5 V)

Figure 6-1. Simple voltage regulator for a 3-volt Hankscraft motor. Powered by a 6-volt battery.

You need only basic soldering skill to make this regulator. The parts can be found at Radio Shack. The wiring can be made on an IC board and tucked into a small experimenter's box screwed into the underside of the platform's bottom leaf (Figure 6-3).

Figure 6-2.

Figure 6-3.

Figure 6-2. Add a voltage meter to a voltage regulator for quick adjustment. Figure 6-3. Small unit attached to the tracker.

A more complex voltage regulator is described by Forrest M. Mims on page 98 of the October 1982 issue of Popular Electronics. His CONTROLLER FOR SMALL DC MOTORS uses VMOS FETS and is designed specifically for low-voltage, low-power motors.

THE HOLBROOK STEPPING-MOTOR DRIVE

Used with a motor control, stepping motors provide performance usually previously obtained only with synchronous motors.

One motor control design is described on pages 80 through 83 in the July, 1986, issue of SKY & TELESCOPE. In this article, John Holbrook discusses what his stepping-motor drive can do and how you can assemble it.

Mr. Holbrook uses a unipolar stepping motor that runs on 5 volts DC and makes 48 steps per revolution.

It is a four-phase type manufactured by Airpax that can be purchased from AST Servo Systems, 115 Main Rd., Montville, N.J. 07045 (Phone 201-335-1007, or toll-free 800-922-11033 outside New Jersey). The stock number is SM-166 and they cost $12.50 each, with a minimum order of three motors.

The wiring diagram and parts list below have been reproduced from SKY & TELESCOPE magazine by permission.

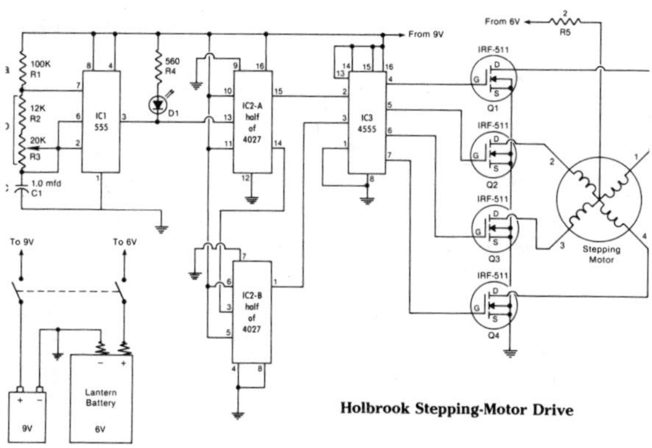

Holbrook Stepping-Motor Drive

PARTS LIST

(Radio Shack stock numbers in parentheses)

B1	9-volt transistor battery (23-583; clip, 270-325)
B2	6-volt lantern battery (23-006)
C1	1-microfarad metal-film capacitor (272-1055)
D1	Red light-emitting diode (276-041)
IC1	LM-555 integrated circuit (276-1723)
IC2	RCA CD-4027-B integrated circuit
IC3	RCA CD-4555-B integrated circuit
M1	5-volt 4-phase stepping motor
Q1-Q4	IRF-511 power MOSFET transistor (276-2072)
R1	100K-ohm resistor, ¼ watt, 5 percent (271-1347)
R2	12K-ohm resistor, ¼ watt, 5 percent
R3	20K-ohm trimmer pot, 15 turns (271-340)
R4	560-ohm resistor, ¼ watt, 5 percent
R5	2-ohm resistor, 2 watts, 10 percent
—	One 8-pin wire-wrap DIP socket (276-1988)
—	Two 16-pin wire-wrap DIP sockets (276-1994)
—	DPDT toggle switch (275-647)
—	Wire-wrap tool and stripper (276-1570)
—	50-foot spool of No. 30 wire (278-501)
—	Perf board with 0.1-inch grid (276-158)

Figure 6-4. Holbrook Stepping-Motor Drive Wiring Diagram and Parts List.

In a letter to me, Mr. Holbrook added some information not included in the SKY & TELESCOPE article:

"The speed stability of my drive circuit is controlled solely by the frequency stability of the 555 timer chip and its associated resistors and capacitors. Typically, this will be in the neighborhood of 100 PPM per degree Celsius over a range of 0 - 70 degrees C. This works out over a 12 degree C range to a worst case time variation of .072 second for 1 revolution."

CONNECTING A HANKSCRAFT MOTOR TO A TURNING BOLT

Cut about 3 1/2 inches from a 10-32 rod. Round one end making sure that the tip is absolutely dead center as the rod turns.

A gear with an 1/8-inch I.D. serves as a convenient connector. On the end opposite the set screw, tap 10-32 threads about one-half way into the gear. Screw the non-pointed end of the rod into the tapped hole. Hold the rod in place with a binder post.

Substitute a thumb screw for the set screw on the gear. This will let you remove the motor very quickly and conveniently (Figure 6-5).

Figure 6-5. A gear makes an excellent connector between the motor shaft and the drive bolt.

Slide the motor shaft into position in the gear. Tighten the thumb screw. This part is now finished. Screw this assembly into the binder post in the lower hinge leaf.

Drill out the pin that connects a lid support arm (a guiding arm for a trunk lid) with the bracket. Replace the pin with a thumb screw and nut. This modification will allow you to replace the motor mount with a timing cylinder you can hand turn (Figure 6-6).

With an angle bracket, a few screws, washers and nuts, you can loosely attach the motor to the guide. After the turning bolt has made a full run, loosen the set screw, remove the motor, and spin the bolt back to the starting position.

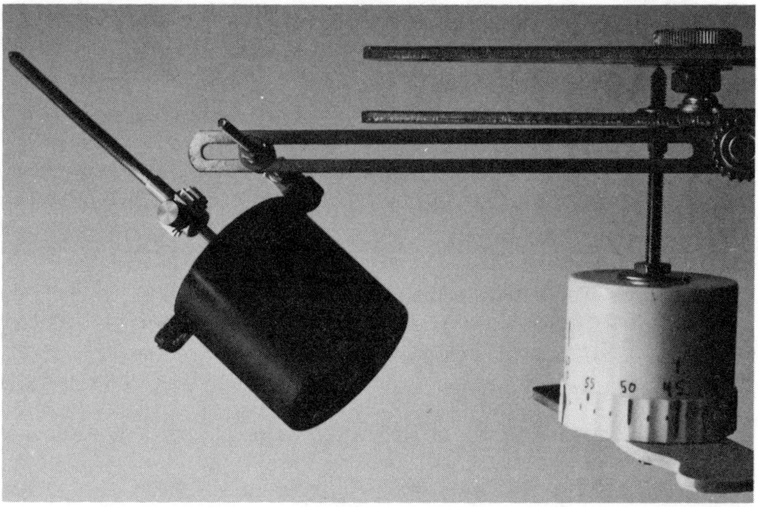

Figure 6-6. Guide can be swiveled to let a hand-driven timing cylinder replace motor.

AUTOMATING A CURVED-BOLT DRIVE

The drive shown on the next page is assembled with the same pulleys, bolt, and O-ring used to build the curved-bolt platform described in an earlier section. This design requires a 10-inch Stanley 902 hinge instead of an 8-inch one.

Just substitute a small DC gearmotor (like the Hankscraft one) for the turning handle. With this design, the upper leaf can be raised or lowered if the motor's gear train is reversible. My 8-inch hand-turned model requires the top leaf to be lowered.

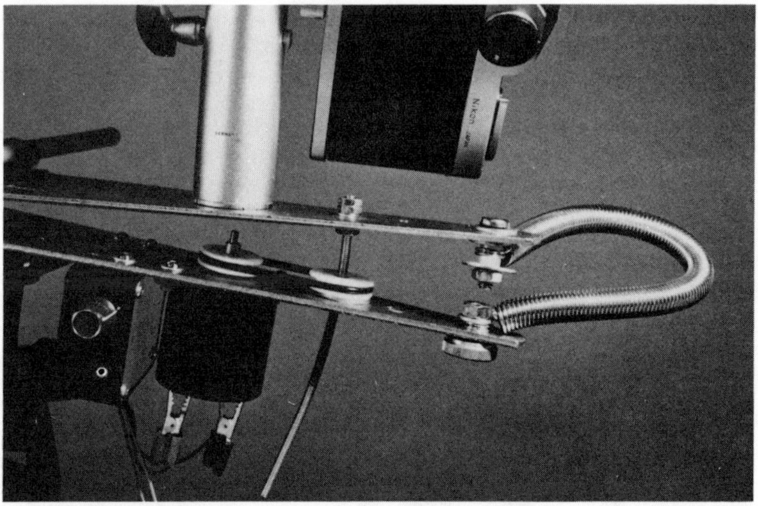

Figure 6-7. An automated curved-bolt drive showing two modifications. Note the friction reducing spring and the voltage regulator.

The motor shaft projects upward through a 3/8-inch hole near the center of the lower hinge leaf. Position the motor and pulley AFTER you've placed the outer pulley and bolt at the required distance from the hinge knuckle position.

A set screw holds the pulley to the motor shaft.

TESTING THE DRIVE

Once your drive is finished, you'll want to test a number of its characteristics. Does it track smoothly? Will it tightly hold the camera/lens combination? What is the longest lens you can use before you start seeing too much movement? Are there ways you can make the platform and tripod support more stable?

You can test either a hand driven or motor driven platform. And, you can "dry" test it inside your house. True, how well the drive tracks the stars is the acid test, but your in-house experiments will give you plenty of useful information.

There are a couple of easy ways to determine how mechanically sound you've made the tracker. One way requires throwing a light beam onto a wall; the other way requires exposing a stationary light source on film as the upper hinge leaf is opened.

Try the "quick test", the first one described below, as soon as you've completed a tracker. You can make the test in minutes, without camera or lens. Use the second method, which requires a camera and lens, when you need a very precise mechanical test result.

A QUICK MECHANICAL TEST

Attach an optical pointer -- the handheld device used to throw an arrow onto a projection screen -- to a straight flash bracket (that's the accessory that screws into the bottom of a camera and holds a flash unit to one side of the camera body).

Mount this combination on a ball-and-socket tripod head atop the tracker to be tested. The projected pointer light will take the place of a camera and lens.

Position the pointer perpendicular to the hinge axis and focus the arrow beam on a wall about 15 feet away.

Tape a piece of paper to this wall area. Lights can be left on for the experiment.

To test, turn the timing cylinder in a series of one-half revolutions, marking on the paper where the beam pointer tip falls each time you stop turning the drive rod. After you've made 15 to 20 marks, connect the points with a line.

An arrow-straight line indicates a sound, mechanical tracker. A wavy line means either a pulley is out-of-round (on a curved-bolt system) or a rod tip is off-center (on a straight-bolt system). Make corrections and continue testing until you get a straight line.

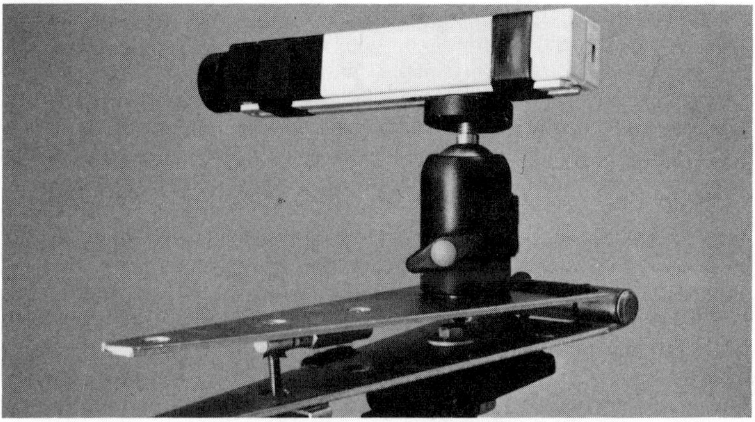

Figure 7-1. An optical pointer mounted atop a hand-driven tracker for quick mechanical testing.

A PRECISION MECHANICAL TEST

Most of this testing can be done in a darkened portion of a house, using a long hallway.

Here's what you do after loading your camera with a roll of high-speed black-and-white or color film:

1) Tape the end of a small flashlight so that only a 1/16-inch square is showing.

2) Place the flashlight about 50 feet from the tracker, making sure the light beam is not aimed directly into the camera lens (it will overexpose the line too much).

3) Set the platform at the approximate angle required for polar alignment. Mount the longest focal length lens you have atop the tracker, turning it 90 degrees from the hinge axis point. This positioning should emphasize any mechanical error.

4) After focusing on the tiny light, start tracking, using times from 4 to 8 minutes, jotting down the settings for each trial.

Use an aperture of about f/11 or f/16. Wider apertures may overexpose and make the shots more difficult to interpret.

READING THE RESULTS

As with the quick test, a thin, extremely straight light trail across the film indicates that the tip of the turning rod in a straight-bolt system is well centered and lifting the upper platform evenly.

A slightly wavy line in a straight-bolt system indicates the bearing point is off-center; a wavy line in a curved-bolt system indicates an out-of-round pulley or an improperly bent drive rod. A hint: count the number of waves in the exposure. If the number of waves equals the number of minutes of exposure, the problem lies in one or both of the pulleys.

This test will dramatically show if a gear-driven tracker has a gear "backlash" problem, a condition caused when the teeth do not mesh tightly enough.

The following illustrations show the results of photographing a point of light with a 500-mm lens atop one of my trackers.

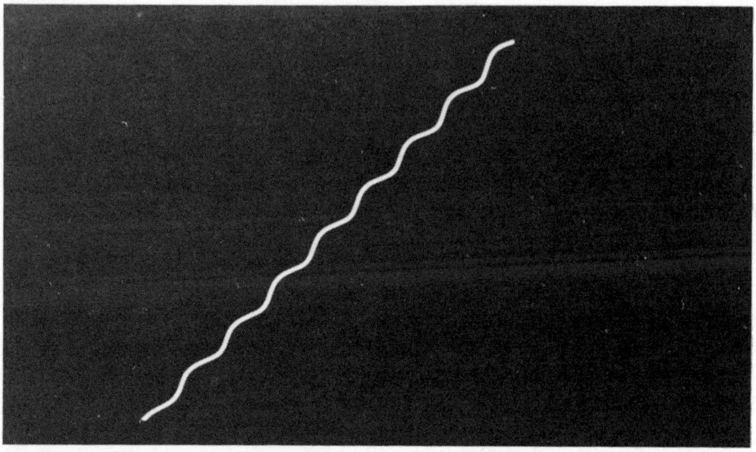

Figure 7-2. This wavy line indicates that the straight-bolt tip is off-center.

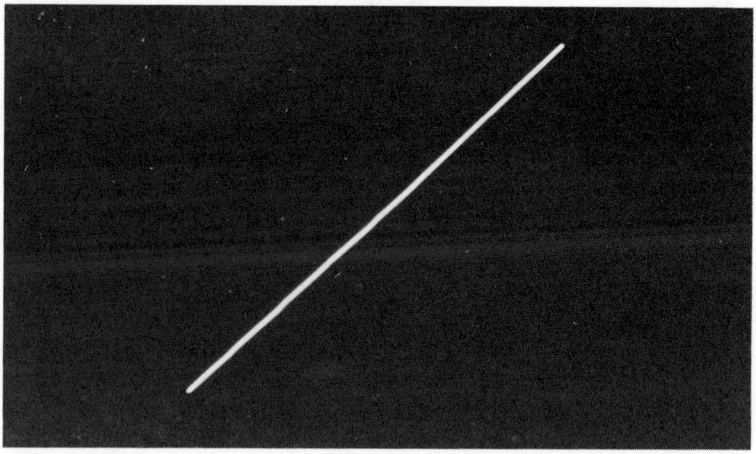

Figure 7-3. The same test as above, but after refiling the turning rod's tip dead center.

A problem might arise when a curved-bolt system's main gear or wheel pulley touches the lower bushing or bearing. If the two surfaces do not remain exactly parallel to one another, uneven lifting -- a stop and go situation -- can occur, possibly creating corkscrew patterns on the film. Adjust the wheel pulley or gear so that it fits exactly into and stays parallel to lower bushing.

By dry testing a tracker, you can modify the design -- refile the bolt point, add more base support, change the way the lens attaches to the upper platform -- until you get an arrow-straight line slicing across a frame of film. Then you know you're ready to turn the unit toward the stars.

FIELD TESTING

A number of factors influence unguided astrophotography from a star tracker. Some of these are specific to the camera, lens, and film -- aspects like lens focal length, aperture setting, film speed and resolution, color sensitivity, and reciprocity failure.

Other factors, specific to the tracker itself, can be tested under actual shooting conditions before an exposure is made. Two of these factors are polar alignment accuracy and drive speed accuracy. Once these crucial conditions prove to be acceptable, your success with astrophotography will be greatly enhanced.

Let's examine polar alignment first. Before you attempt to polar align a mount, you must precisely align the finder to the hinge axis. Do this by positioning the finder on the top leaf parallel to the hinge line. The finder is aligned accurately when a distant object, placed dead center in the finder, remains centered as the upper leaf is opened as far as possible.

POLAR ALIGNMENT

In the Northern Hemisphere, true celestial north (Figure 8-1) lies about 0.8 degrees away from Polaris toward the star Mizar in the Big Dipper.

In the Southern Hemisphere, the pole star corresponding to Polaris is the 5th magnitude star Sigma in the constellation Octans. Sigma is located approximately 1 degree in the 21hr direction from the South celestial pole (Figure 8-2).

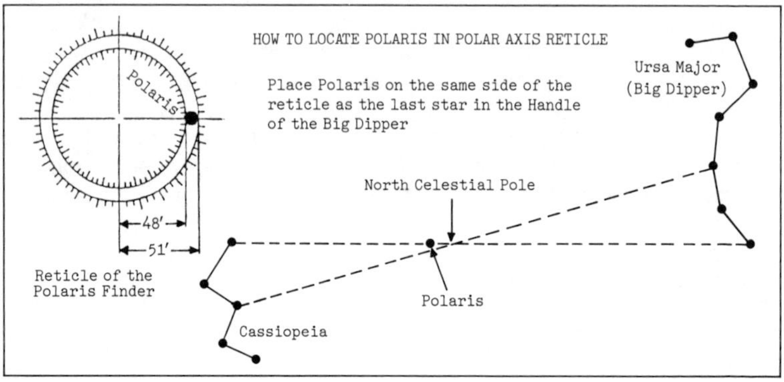

Figure 8-1. True celestial north.

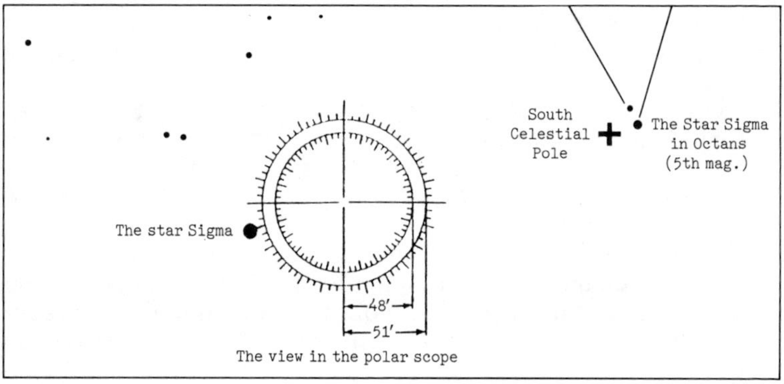

Figure 8-2. True celestial south.

For this test, you'll need a stop watch and either a finderscope or riflescope with crosshairs. Use a red-filtered flashlight angled into the front of the scope to illuminate the reticle. You won't need a camera.

Mount the scope on the upper leaf's ball-and-socket head. After you have polar aligned the tracker, position the scope so that one of the reticles is parallel to the horizon. Fix the center of the scope's crosshairs on a prominent star at least 30 degrees over the eastern horizon and about 90 degrees from the polar star.

If you are testing a hand-driven 1 r.p.m. tracker, turn the bolt 15 revolutions. Start the stop watch. If testing an automated drive, switch on the motor. At precisely 15 minutes, look through the scope. The star should have "caught up" to the crosshairs and be in its original starting position in the scope. If it is, run the test again without making any changes.

If the target star has crossed over the VERTICAL crosshair and now lies more north or south of the original position, you'll have to adjust the polar alignment. If the star lies above or below the HORIZONTAL crosshair, you'll need to adjust the tracking speed.

Since high-powered rifle scopes show the image right side up and finderscopes show things upside down, right to left reversed, you will have to adjust for polar alignment accordingly.

Keep testing, making minor changes. When the star remains fixed in the center of the crosshairs for 30 minutes, note where the pole star is situated in the finder. If you had to slow or speed up the drive, note exactly how many seconds change you made.

Run this test, even for shorter times, when you need to be absolutely certain of alignment and timing. Information gleaned from this "no-camera" testing can greatly improve your results.

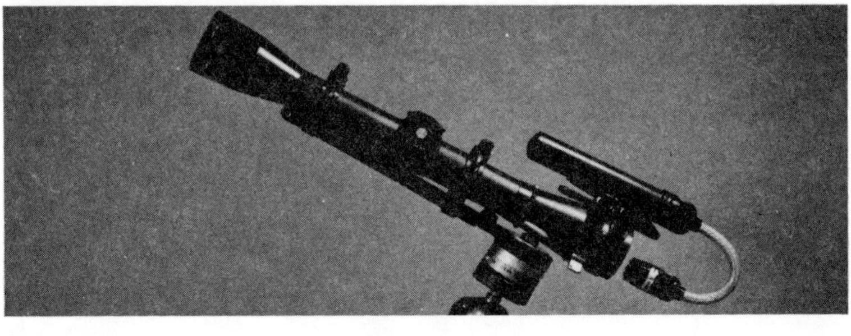

Figure 8-3. Illuminate the crosshairs of a riflescope by slanting a red-filtered light into the back optics. This type of flexible neck lamp works well.

SHOOTING WITH THE TRACKER

Now comes the joy of trying out your new tracker in a real situation, under the stars. If possible, choose a time when the moon is new, or at least, not at its brightest. (Subscribe to the SKY CALENDAR, a monthly single-sheet sky chart for exact moon phases, star positions and other planetary information. A year's subscription costs only $6, one of the best buys around. Write to SKY CALENDAR, Abrams Planetarium, Michigan State University, East Lansing, MI 48824.)

There are a few things you should do before arriving at the site: Load the camera (See section on Film in Appendix) and shoot at least three frames of a lampshade or some other light subject. These throw-away frames will give the film processor cutting mark guides.

Also, remove the camera batteries if the shutter can be activated independently. Long exposures quickly drain camera batteries. See the section on BATTERIES for details on making an alternative external battery chamber.

Be sure to take along a cable release, notebook, star guides or charts, and extra flashlights.

THE SITE

Get as far away from street lights and city glare as you can. Sky fog from mercury vapor or sodium vapor lights can severely limit exposure times, reducing the number of stars the film will record. Find an open field away from the tops of trees -- an open field that will give you nearly 180 degrees of view is perfect. A call to a local astronomy club should give directions to the closest dark, accessible places.

At the shooting site, set up your tracker using a red-filtered flashlight. The red color will not curtail night vision, necessary for observing fainter stars and deep-sky objects. While at the site, use only red-filtered lights for note-taking, reading star charts, or polar aligning the tracker.

A tip for polar alignment: after getting the mount roughly aligned to a celestial pole, nudge the tripod legs for fine alignment. This will work better than adjusting the pan/tilt head.

CAMERA ADJUSTMENTS

Set the camera's focus at infinity, the shutter speed to the B setting, and screw a cable release into the shutter release. Set the aperture to f/2 (if possible) for edge-to-edge star sharpness, or even to f/1.4 or f/1.2 if some loss of edge sharpness is not important.

Turn the camera skyward, framing the star or stars you want to capture. Keep important stars as close to the center as possible for greater image resolution. A right-angle viewer may make focusing and composition easier.

Lock the camera in place. Attach a red light to the underside of the tracker and aim the beam so that it strikes the clock surface and also strikes the timing cylinder.

If possible, lock the camera's mirror up. This will greatly eliminate vibration. Finally, make sure that both the camera and the base tripod legs are securely tightened before you take the first exposure.

MAKING THE EXPOSURE

Hold a black card as close as possible in front of the lens without bumping it. Hit the cable release. Wait a few seconds before lowering the card. The card, in essence, begins the exposure.

The instant you remove the card from the lens, begin slowly and smoothly turning the handle and timing cylinder. Match the timing cylinder seconds to the seconds on the clock and do this second by second until the full exposure time has elapsed.

You will develop a rhythm so that even when you glance away from the clock your fingers will keep moving at exactly the right speed. Fellow tracker Frank Zullo eloquently describes this time as a way "of keeping pace with the Earth's rotation by just the swing of one finger." He considers this method of timing "probably the most exciting thing about these trackers."

CHECKING THE SKY

Periodically glance up to the sky where you are shooting. Is it still dark and clear? Are there any clouds threatening to block the star light?

If a plane, with all its flashing lights, is heading into your "piece of sky," just block the lens with the black card again, wait for the plane to pass while continuing to "drive" the tracker, and then resume the exposure. You might be able to block the light from satellites, but don't count on it. There are large numbers of them, and with their speed, they are hard to prepare for. Their long trails will soon be a common, though visually unwanted, feature in your star shots.

At the end of the exposure, block the lens with the black card BEFORE you reach for the cable release. This will prevent any additional vibration from affecting the shot. It will also eliminate any problems if you accidentally bump the camera or tripod as you release the shutter.

TAKING NOTES

Immediately write down the details of the shot: the date, time, star or constellation, f-stop, exposure time AND whether the shot was taken horizontally or vertically. When using short exposures on new portions of the sky, don't trust your memory too much. A "V" or "H" marked next to the frame number will identify the orientation.

Sketch the position of a prominent star in the frame if you're using a long telephoto and that star is placed in one corner of the frame. Fainter, nearby stars won't help you identify this subject in the frame, but your sketch certainly will.

Occasionally, you should re-check the polar alignment. Is the pole star in the same relative position as when you started? Perhaps the tripod leg slipped a little, or the platform sagged, or ... who knows what can happen. It takes only a few seconds to realign the tracker if it is misaligned, but those few seconds can prevent hours of fruitless exposure time.

REPEATING AN EXPOSURE SERIES

Use a repeatable exposure time series. For example, take shots for 1 minute, 3 minutes, 6 minutes, 9 minutes, 12 minutes, 15 minutes, and so on. By repeating this series, or one of your choosing, you will know at a glance the exposure times of your slides or negatives, even if you haven't taken notes.

Film reciprocity failure -- when the emulsion performs less efficiently during long exposures in faint light -- often requires that you triple, rather than double, the exposure time for each consecutive shot to capture twice the amount of light (opening the aperture one f-stop). This general rule will apply to many films in the 200 ISO to 400 ISO range, but study the films' spec sheets for specific information.

If you have unexposed film in the camera after an evening's shooting, you can remove it safely. Backwind it, making sure the leader doesn't wind back into its cassette. Mark on the side of the film how many shots were taken and the date, replace it in a plastic film canister, secure it in a sealed plastic bag and store it in the freezer.

To reuse the roll, let the film sit at room temperature for about 6 hours, then reload into the camera. Just cover the lens with a lens cap and your hand and "reshoot" the frames (actually reshooting blackness) until you get to the unused, unexposed film. You're ready to shoot again.

You should remove the film and put it in cold storage whenever the film is hypersensitized or has an emulsion that is affected by temperature changes. Film spec sheets will give you more details about storage and temperature concerns.

If you reload the film, shoot a few frames of a lampshade at the beginning of each new section.

Once the film is ready for processing, tell the processor there are star images on the film. Too often the processor will mistake those hard-earned star shots as "light leaks and pinholes" not deserving of mounting or printing. Once, even after telling a processor what was on the film, he brought me the entire roll, uncut, and said "Nothing came out. Should I just toss it?" Close inspection showed that the frames were filled with hundreds of stars! After that experience, I've developed every roll of star images I've taken. I suggest you do the same.

IN THE CLEARING

Watching the small, illuminated clock dial, I turn the tracker's handle ever so slowly. One second, two seconds, three seconds ... the hinge opens ... twelve seconds, thirteen seconds ... reading the always changing clock and matching the timing mark with the clock number ... twenty seconds, twenty five ... moving my wrists and fingers as smoothly as possible to reduce vibration to the camera ... fifty-five, sixty, "click." The one minute exposure will bring out stars my eyes can't see. One minute is a very long time experienced second by second.

I start again. During the next exposure I hear, but can't see, a deer moving through an alder stand a couple of hundred feet in front of me. As I move the turning screw, the deer moves toward me, pauses, looks, bends down to feed, looks again, and waits. I hear what he is doing, know how he is watching.

Sitting there, I think of how, two weeks ago, I watched a porpoise come toward my kayak. He had seen me from some distance away and had turned to investigate. I stopped paddling and let the kayak come to a stop on the mirror surface of the ocean. The porpoise submerged and remained out of sight for some time. Then, he resurfaced four feet from me, sleek, quiet, bringing with him a beauty beyond words. He saw me and knew I was watching him.

The turning screw forces me to slow down, forces me to slow down and repeat the same gesture. I move so precisely for so long that I move into a trance, a kind of aware meditation. I move with a regular, circular motion. To speed up or slow down will cause the stars to blur. I keep going.

I remember sitting this quietly in a forest during a snowstorm. I just wanted to be out in it, discovering what would happen. Snow piled up on my legs and arms and shoulders. The day was so white I couldn't tell the ground from the sky. Then I heard birds singing, coming closer. Chickadees. Out feeding or just playing in the storm. They hopped from limb to limb above me. One of them flew down and landed on my hat. Then another. They hopped about my head, kicking up snow, singing loudly in my ears.

I come back to my star tracking. Even as I drift into memories, my turnings hold true. The seconds match. I keep turning to the earth's speed.

Sitting here beneath the night sky is as magical as the way a deer steps through the woods, as magical as a porpoise surfacing, as magical as chickadees dancing in a snowstorm.

I keep going. I move my hand in quiet concert with the stars.

Figure 9-1. A straight-bolt metal tracker was used to record the star Sirius (upper right) and open star cluster M-41 just below and to the right. Original shot was an 8-minute exposure on Fujichrome 1600 D with a 105-mm lens set at f/2.5.

Figure 9-2. A 6-second exposure of Deneb (the brightest star) and the North America Nebula region. This shows the approximate number of stars visible to the naked eye when the observer is near lights from a large city. A stationary shot taken with a 105-mm lens on Ektachrome 400 film at f/2.5.

Figure 9-3. Even a short, 3-minute tracking shot of the same Deneb region shows the increased number of stars that can be recorded on Ektachrome 400 film using a 105-mm lens at f/2.5.

Figure 9-4. A 5-minute non-tracking shot of the same Deneb region. Taken with a 105-mm lens at f/2.5.

Figure 9-5. A 5-minute tracking shot of the same region in Figure 9-4. Below Deneb, at the right-center of the photo, the North America Nebula is visible.

Figure 9-6. Cygnus region of the Milky Way. A 9-minute exposure on Fujichrome 1600 D film using a 24-mm lens at f/2.8. Made with a curved-bolt hand-driven tracker.

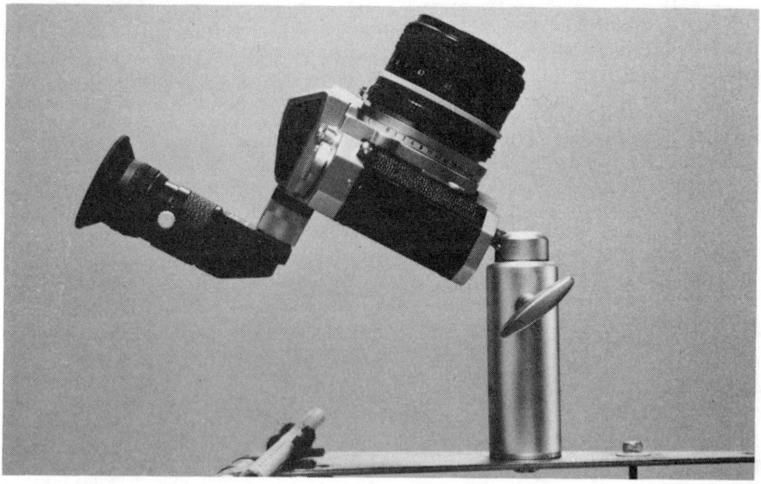

Figure 9-7. A right-angle finder allows you to easily compose star shots when the camera is steeply tilted over the hinge line.

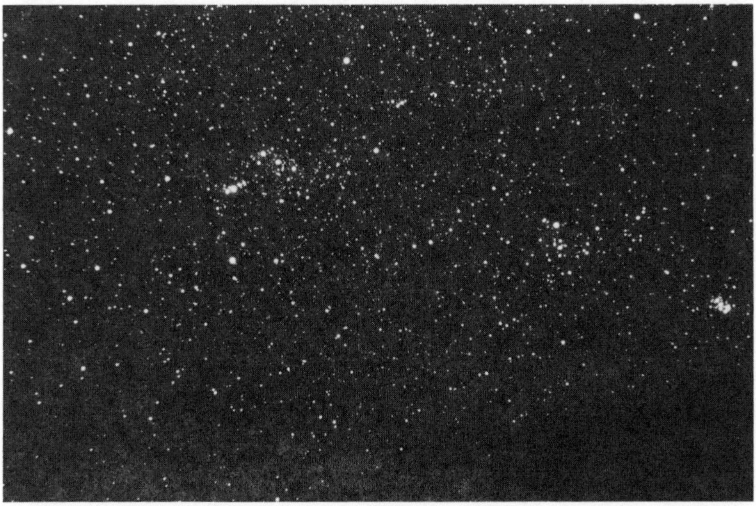

Figure 9-8. A 24-mm lens gives a wide-sky view of the Orion, Taurus and Pleiades region. Galaxies, constellations, and deep-sky objects can easily be framed and seen in context when wide-angle lenses are used. Original shot was a 9-minute exposure on Fujichrome 1600 D at f/2.8.

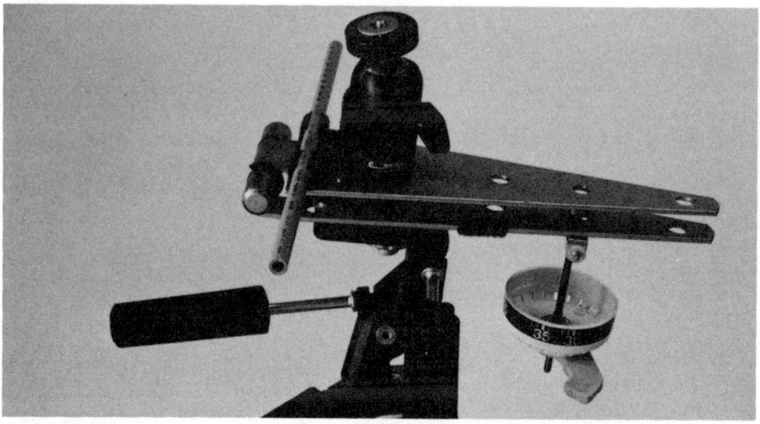

Figure 9-9.

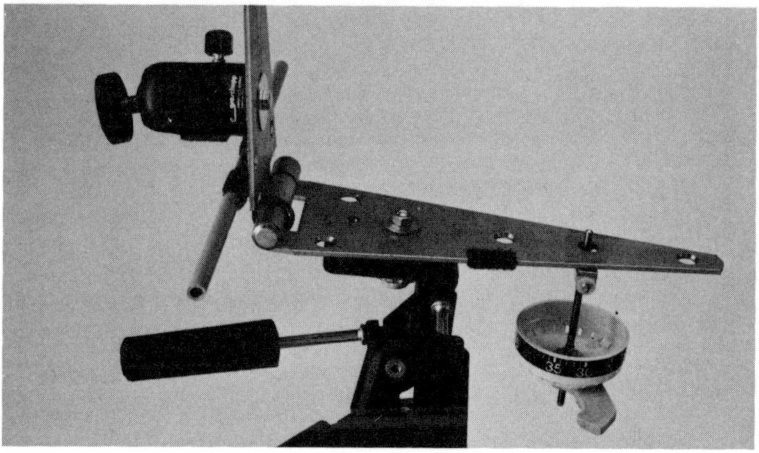

Figure 9-10.

One of the most important aspects of setting up the tracker is verifying that the polar alignment sighter is exactly parallel to the hinge axis. A subject centered in the sighter should remain in that position when the hinge is closed (Figure 9-9) and when opened 90 degrees or more (Figure 9-10).

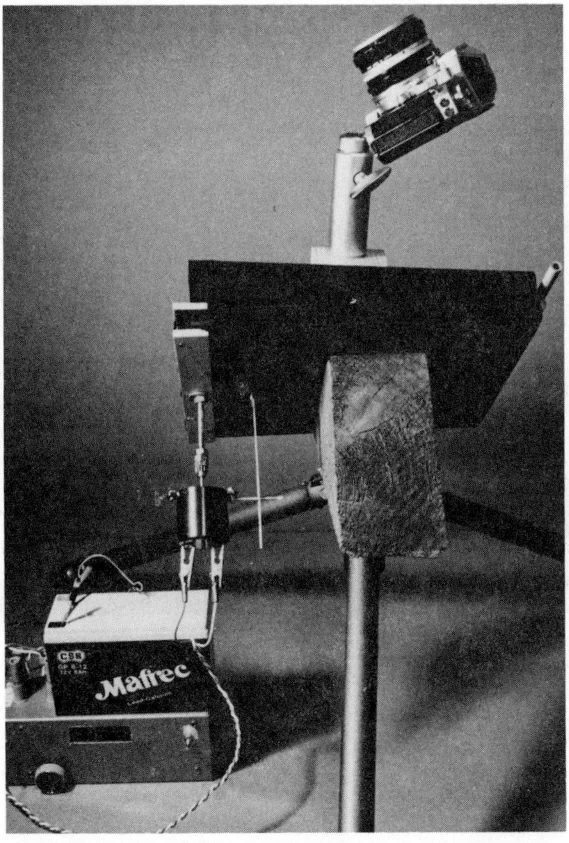

Figure 9-11. An automated straight-bolt wooden tracker. Its 1 r.p.m. Hankscraft gearmotor is powered by a 12-volt lead-calcium battery and an experimental motor control unit still being field tested.

The motor assembly can be removed and a hand-driven timing cylinder substituted if desired.

Figure 9-12. An automated curved-bolt metal star tracker was used to make this 12-minute exposure of the Andromeda Galaxy. Taken with a 105-mm lens at f/2.5 on hypered Kodak Tech Pan 2415 film.

APPENDIX A

CONDITIONS AFFECTING UNGUIDED ASTROPHOTOGRAPHY

THE UNGUIDED SHOT

If you are not altering the speed or making minute adjustments for direction during an exposure, you are taking an UNGUIDED shot. A number of factors influence the outcome of unguided astrophotography. Here are some things you must consider. These points are also thoroughly discussed in the fine book ASTROPHOTOGRAPHY by Barry Gordon, published by Willmann-Bell, Inc.

POLAR ALIGNMENT. The closer you can come to getting the tracker perfectly polar aligned, the longer your exposures can be before star-trailing sets in.

ATMOSPHERIC STEADINESS. The light from a star near the horizon passes through more atmosphere than light from a star at the zenith, the sky directly overhead. This increase in atmosphere makes the star appear to move more slowly going upward from the Eastern horizon, speed up as it approaches and moves through the zenith, and slow down again as it nears the Western horizon. This is called DIFFERENTIAL REFRACTION, and it won't affect short exposures using wide-angle or normal lenses. But long exposures with long (over 135 mm in focal length) lenses could be affected.

PLATFORM STABILITY. A stable platform will hold the camera and lens rock-solidly as it moves. Hinges with too much play, a wobbly tripod, a platform either too thin or with too little surface area, a drive with either periodic or random mechanical problems are a few conditions that will rob your efforts at getting first-rate star shots. It's extremely important to build your tracker to overcome all of these potential problems.

FOCAL LENGTH OF THE LENSES. Wide angle to normal lenses can usually track accurately up to 15 minutes with no guiding. A stable platform, precise polar alignment and accurate drive speed can increase that time to 25 to 30 minutes.

Telephotos (any lens above 50 mm focal length) are much harder to use successfully. Because such a small area is taken in through such lenses -- about 2 degrees with a 1000-mm lens -- even the slightest vibration in the drive system is transmitted to the lens which results in "worm trails", corkscrews or spikes.

Learn to use wide-angle and normal lenses until you consistently get non-trailing star shots. Then work up to longer lenses, starting with short times and gradually increasing them. Trials and many errors will dictate how many minutes you can use with that super telephoto. Making a difficult exposure -- say 4 minutes with a 500-mm lens -- and coming out with crisp star images is a wonderful reward for the many hours spent star tracking.

YOUR OWN PERSONAL IMAGE QUALITY. You're the final judge when it comes to defining what makes a good star image. A small amount of trailing, rendering a star the tiniest bit egg-shaped, might be perfectly acceptable to some of you but frowned on by others.

APPENDIX B

FILMS FOR ASTROPHOTOGRAPHY

The size of the target subject, viewing conditions, the focal length and aperture of the lens, the position of the star and other factors should influence your choice of film. This is only a starting point.

TRANSPARENCY FILMS

Fujichrome 400 D and 1600 D
Ektachrome 400

COLOR NEGATIVE FILMS

Kodacolor VRG 200, VRG 400 and VRG 1000
Konica SR 400, SR 1600 and SR-V 3200

BLACK AND WHITE FILMS

Kodak Tri-X (400 ISO)
Ilford XP1 (400 ISO)
Kodak Technical Pan 2415 (Hypered for greater speed)

****

In a letter to me, Michael Covington, author of ASTROPHOTOGRAPHY FOR THE AMATEUR, cited a few other specific films:

"3M ASA 400 color slide film (marketed as K-Mart "Focal" 400 film in the U.S. and as a house brand of the Boots pharmacy chain in England) has markedly less reciprocity failure than Ektachrome 400. It is also a bit coarser-grained. I like it for sky exposures in the 1- to 10-minute range.

"Kodak T-Max 400 and T-Max 100 films will have markedly less reciprocity failure than Tri-X or comparable films. Early Kodak technical data sheets show data that translate to a Schwarzschild exponent of about 0.80 for both films (as compared to 0.6 for most ISO 400 films). This means that T-Max 400 will be faster than hypered 2415 in exposures of several minutes' duration, and T-Max 100 will be almost as fast as hypered 2415. At last we will have 'off-the-shelf' films that are good for deep-sky photography."

Dennis di Cicco added that "T-Max 100 and T-Max 400 are good films, but both lack extended blue and red sensitivity."

APPENDIX C

STRAIGHT-BOLT VS. CURVED-BOLT DESIGN

The diagram below, drawn to scale using a 32 t.p.i. drive rod, shows some differences between a straight-bolt hinge drive and a curved-bolt drive.

Notice that both the straight-bolt (vertical line) and curved-bolt (curved line) remain fairly close together during the first 1/2- to 3/4-inch (or 16 to 24 minutes). Beyond 1 inch (32 minutes) the two systems deviate substantially.

This illustrates why a straight-bolt system can closely duplicate the results of a curved-bolt system at times approaching 25 minutes. Beyond that time, the straight-bolt must be turned faster to compensate for its tangent error.

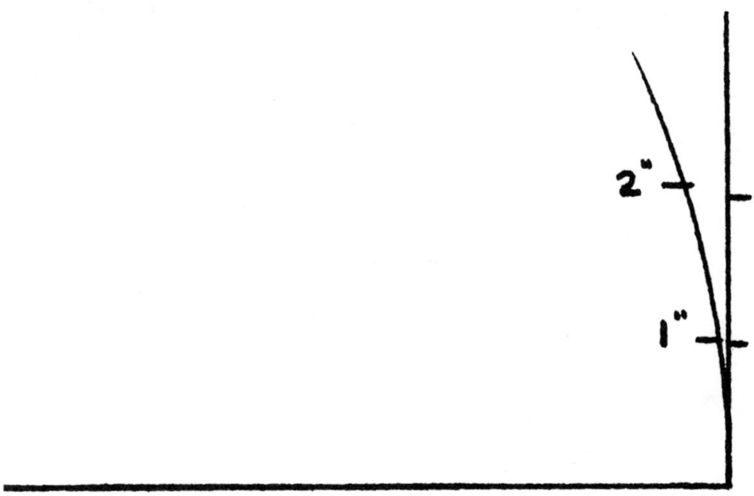

Figure A. Comparison of straight-bolt and curved-bolt drive systems.

Regardless of whether you use a straight-bolt or curved-bolt tracker, it is a good idea to start a series of photographs with one stationary shot. This will serve as a reference image, one that will show you the target star, constellation or deep-sky object at a glance. A 10- to 15-minute shot may include so many points of starlight that "reference" stars are hard to find.

Use the formula below to help you determine how many seconds of exposure you can use before star trailing sets in. This is a rough formula, so you should experiment and make adjustments if necessary. The formula is for photographing a star on the celestial equator.

$$\text{Time In Seconds} = 600/F$$

In this equation, F is the focal length of the lens being used. The table below shows a number of commonly used focal length lenses and maximum exposure times possible before star trailing occurs.

FOCAL LENGTH OF LENS	MAXIMUM PERMISSIBLE STATIONARY EXPOSURE TIME (SECONDS)
20-mm	30
24-mm	25
28-mm	21.4
35-mm	17.1
50-mm	12
80-mm	7.5
100-mm	6
135-mm	4.4
200-mm	3
300-mm	2
500-mm	1.2

APPENDIX D

COMMERCIALLY AVAILABLE STAR TRACKERS

This section discusses a few commercially available tracking systems and gives addresses where you can write for more information and current prices.

In some cases I've included conversations I've had with the managers, designers or distributors of the trackers. This may give you a clearer picture of some not-in-print aspects of the trackers.

THE SKY PATROL

"I use my Nikkor 300-mm f2.8 on the Sky Patrol," says Art Ciampi, Marketing Manager for the Texas Nautical Repair Co., distributor for the Takahashi Sky Patrol. "You can mount any large cat lens or even a C90 spotting scope with no problems.

"The mount will hold about 6 pounds. The unit, with camera and lens, usually weighs about 15 pounds total. It comes with a carrying bag that's about the size of a woman's large handbag. It's very portable. A lot of people bought this tracker for shooting Halley's Comet. We quickly sold 250 units, and I know we could have sold 500 if we had them in stock. Once people see them, they're gone. In fact (addressing me directly)... once YOU see one of the Sky Patrols, you're a dead man. You'll have to get one.

"I just returned from a trip to Tokyo where I watched the Takahashi workers hand assemble each unit. A roomful of women were painstakingly assembling the eyepieces. They were very meticulous and precise. It was really impressive.

"Built-in grids give very accurate tracking when using the guide scope. I might adjust the drive slightly every half-minute. I can hear the tracker motor's putt .. putt .. putt .. as it smoothly keeps the star centered in the crosshair.

"The Sky Patrol is actually a very stable equatorial mount with built-in slow motion drive correction control."

Get Art Ciampi talking about the Sky Patrol and you quickly hear superlatives that have no match. Art's a man happy with the product he sells.

He's not alone with this praise. Dennis di Cicco writes this about the mount: "A unit similar to the discontinued Japanese pulse-motor drive made by Vixen (in some ways better, since it has provision for corrections in declination) is the Takahashi Sky Patrol sold by Texas Nautical (ads in SKY & TELESCOPE). The machining is a bit better than my Vixen, but it also costs more -- about $600. I give my highest recommendation to these camera mounts."

Figure B. Takahashi Sky Patrol. Photo courtesy of Takahashi Systems.

Designed and manufactured by Takahashi Systems of Tokyo and distributed in the United States by the Texas Nautical Repair Company, the Sky Patrol is a portable equatorial mount that can be used for both visual observations as well as deep-sky photography with a camera and lens combination. In addition, the mount can be used with a small aperture refractor -- like their FC-50, a 50-mm diameter, f/8 fluorite apochromat -- to achieve extremely sharp, high contrast images.

Takahashi's description of the mount: The built-in polar alignment finder scope is provided to allow easy polar alignment to within .01 degree of the celestial pole. The declination slow motion makes it easy to bring a guide star or celestial object into the center of the field of view.

A piggyback plate with a slit is mounted on the declination axis of the mounting, and a large knob is provided which will allow the mounting of a camera/lens combination, small refractor (FC-50), or a camera pan head. The Sky Patrol unit may be mounted on a standard camera tripod using the 1/4-by-20 threaded hole on the base plate, or the optionally available wooden tripod. The automatic tracking HD-4 precision quartz motor control is included. The HD-4 motor control is also a drive corrector for guided astrophotography. In addition, there is a slow motion motor control included. Using the motor drive, deep sky astrophotography with lenses of up to 200 mm can be be made for up to 10 minutes without any guiding.

Other important information if you want to guide with this tracker: The guide scope with an effective aperture of 40 mm and a focal length of 240 mm is available. With the optional 2x illuminated barlow lens, the guide will then have an effective focal length of 480 mm, and with the 3x illuminated barlow it becomes 720 mm. The guide scope is adjustable in all directions as attached to its arm. There are two threaded holes provided for mounting a 5 X 24 finder scope if needed.

The Sky Patrol Guide Scope specifications:

Effective aperture .... 40 mm
Focal length .... 240 mm
Resolving power .... 2.9"
Light gathering power 32
Limiting magnitude 9.8

To use the 40-mm as a guide scope, you must purchase the 2x illuminated barlow in addition to the basic Sky Patrol mount. The current price of the Sky is listed below.

The equatorial camera/telescope mount with a 40-mm guide (less the 2x barlow), motor control, drive corrector is about $550. The Sky Patrol is available without the guide scope for about $440 for those making unguided shots only. The 2x illuminated barlow costs about $125.

For more information, write or call:

Texas Nautical Repair Co.
American Optics Center
2129 Westheimer
Houston, TX 77098
Phone (713) 529-3551
(713) 529-8480

****

THE CLOCKWORK ROTATION MECHANISM

The name of this star tracker pretty much explains how it works. A fairly compact unit (about 2.6 pounds), the tracker uses a key-wound clock drive that turns a shaft at a constant, precise rate. By using a 5 x 24 finderscope that is positioned parallel to this shaft, the mount can be polar aligned. The camera and lens are attached to a bracket that extends perpendicular to the shaft; on the opposite side of the shaft, a sliding weight can be adjusted along a metal rod to counterbalance the camera equipment.

"This drive operates at the sidereal rate," said Wil Achterberg, spokeswoman at Purivox, the U.S. distributor and agent for the clockwork mechanism. It's quite accurate.

"When wound with the kind of key you use on a Grandfather's clock" she said, "it will continue running for seven to ten days."

She added, "It should be wound a few times each shooting session you have, even if the clock is still running. Be sure to wait a few minutes, however, before you take the exposure."

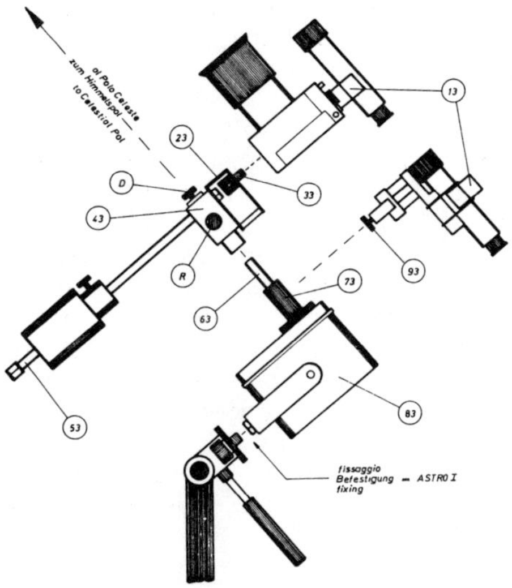

Figure C. Exploded view of the Clockwork Rotation Mechanism. Illustration courtesy of PURIVOX.

Some specifics: You use eight to ten half-turns of the key in a clockwise position to start the tracker. You must wait at least one minute for the clock to stabilize so that your exposures will be consistent over time. The clock drive cannot be interrupted once it's set in motion.

If you decide to shoot a day after winding the drive, give the clock another three half-winds to insure consistency. In other words, even though the clock will go on and on, "restart" the drive before each shooting.

How much weight will the drive hold? "It's best if the combined camera and lens weight be 4 pounds or less," said Wil Achterberg. "You could go heavier, but we don't recommend it. More weight would require heavier counter-balances."

The drive's accuracy? The manufacturer claims the accuracy of the right ascension axis is ± 15 arc seconds per hour.

The drive is designed for Northern Hemisphere shooting, but can be supplied with a clock whose drive runs counter-clockwise. You can specify the exact requirements.

Unlike the Sky Patrol mount, this clockwork system is not designed for use with small telescopes. The distributor says that the clock drive will not give proper right ascension motion unless exactly counter-weighted from the hour axis of the telescope mount. The manufacturer makes no claims for maximum exposure times with various lenses. They suggest that you use the trial-and-error method of testing, eventually working up to longer and longer telephotos.

The clock drive is not dependent on batteries or any external power source. It's self-contained and requires no cords or extra components to make it work. It's light and portable.

The tracker can be purchased with or without their tripod.

Model 1 (Tracker less tripod)	$239.50
Handling and shipping	5.75

Total	$245.25
Model 2 (Tracker with tripod)	$324.50
Handling and shipping	12.50

Total	$337.00

To order or to get more information:

PURIVOX
Evert Achterberg
P.O. Box 123
Escalon, CA 95320
Phone (209) 838-1259

THE KAJERS 7700 ASTRONOMICAL CAMERA DRIVE

"A friend of mine, Ed Roberts, designed this drive. When he died five or six years ago, I continued manufacturing them." Roy Thomas was describing how the Kajers 7700 Astronomical camera drive came into being. "Ed designed the basic drive, and I just added a few things here and there."

Figure D. The Kajers 7700 Astronomical Camera Drive weighs less than 3 1/2 pounds and is crafted from machined aluminum alloy castings, stainless steel, and bronze.

This compact (4 x 6 x 8 inches) star tracker uses a precision two-stage worm reduction drive. Some of its other features include:

1) A high-torque permanent magnet synchronous motor to drive the gear train at the sidereal rate when operated directly from the power line; the power cord is included.

2) A graduated declination scale.

3) An hour circle that allows the camera to be set at the Right Ascension position of any object in the sky.

4) Provision for mounting the camera drive directly on your tripod, using the standard 1/4"-20 screw thread.

5) An optional right angle North Star Finder scope for precise polar alignment.

How much weight will the mount hold? Mr. Thomas said the drive will easily handle from 5 to 7 pounds.

How long of an exposure can be made with various lenses?

"The main worm gear makes a complete cycle every four minutes," he said. "If there is going to be an error in tracking, it will repeat itself at this interval." He went on to say, quite correctly, that a great deal of the accuracy depended on how well the drive is polar aligned.

Mr. Thomas said that their 7500 Telescope Drive/Corrector can be used with the Camera mount -- and a guide scope -- for taking longer exposures.

The star tracker uses the same Right Ascension worm drive that Kajers uses for larger telescope mounts. Their 300 inch-ounce 10 r.p.m. permanent magnet synchronous motor (it has 50 times the torque output of the timing motors supplied in the C5 and C8 Celestron telescopes) drives a precision worm gear train which in turn rotates the Right Ascension axis at the precise rate of 1,436 minutes per revolution. By operating this motor at frequencies other than the line frequency, any R.A. tracking rate -- including solar and lunar -- can readily be achieved.

Since this tracker requires a 115 V, 60 Hz power source, you will need to buy a portable AC-DC power converter for field work. These converters, which sometimes include drive correctors, are available from KAJERS, Orion, Optica B/C Company and other astronomy equipment suppliers. You can power the drive from a stand-alone battery or from the cigarette-lighter jack of your car.

An optional switch can be ordered that will reverse the direction of the motor. This will allow the drive to be used in the Southern Hemisphere. Expect to pay about $70 extra for this North/South switchbox.

The 5 x 24-mm right angle finder scope is optional and costs about $44, including the mounting brackets.

The KAJERS 7700 Astronomical Camera Drive costs about $269. For operating at 230V, 50Hz, add $12. Shipping costs will not be more than $8.

To get more information or to order:

KAJERS
1112 W. Vine St.
Springfield, IL 62708
Phone (217) 787-0861

****

THE ASTRO KITS CURVED-BOLT DRIVE

Using a "hinge" style platform, Astro Kits offers a compact, easy-to-assemble curved-bolt drive. It has a number of interesting features that are discussed below.

To assemble the kit, you'll need to have basic soldering skills; the parts go together quickly and easily, and the instructions are thorough. If you do run into any problems, you are told to call their number -- day or night -- for extra help. If they're not in, leave a message on the answering machine and Dennis Harper, Astro Kits' manufacturer and distributor, will call back.

"To date," Harper says, "no one has called wanting additional information on constructing the kits."

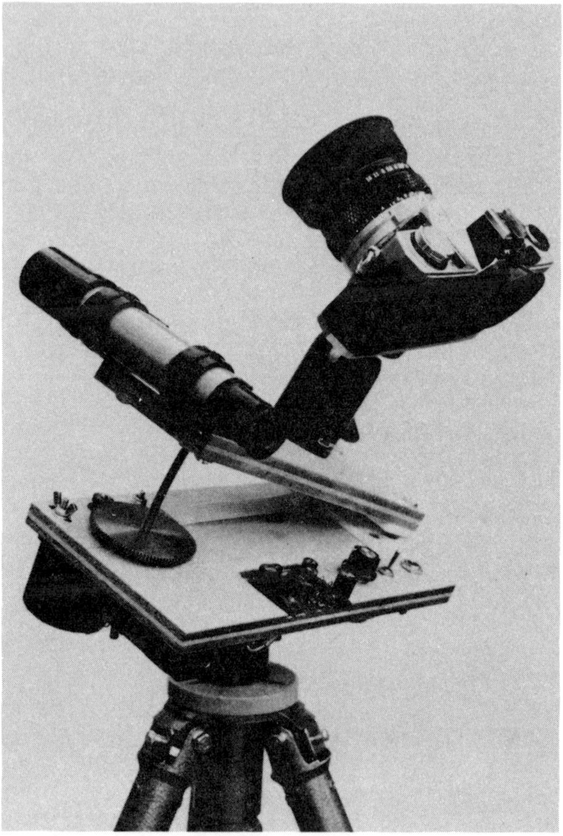

Figure E. Astro Kits curved-bolt drive.

The Astro Kit drive uses the curved-bolt method of controlling the rate of descent of the upper deck of the hinge drive. This method eliminates tangent error, an unwanted characteristic of straight-bolt drives.

The drive uses a 32 threads-per-inch brass rod that is situated the required 7.14 inches from the hinge axis.

The Astro Kit is driven by a motor that has a no-load speed of 5.5 r.p.m. with a nominal stall torque of 15.4 inch-ounce. The internal gear reduction is 342:1.

An 18-tooth all brass gear, mounted on the motor shaft, meshes with a 100-tooth main gear. The hub of the 100-tooth gear is shaped to allow it to seat easily and snugly into its bearing. Once seated with the camera mounted on the upper deck, there is no side play. The hub has a shoulder to prevent it rubbing with the face of the bearing or the lower deck.

The center of the bearing is a tapped 32 t.p.i. hole through which the curved rod moves very smoothly. With this system, the main gear must rotate precisely 1 r.p.m. to achieve the correct upper-deck-descent rate. A spot of red nail polish at the edge of the main gear serves as a timing mark for fine adjustments to the speed.

The rod is long enough to allow about 2 hours of continuous shooting, far more time than you would ordinarily require. The rod is precisely bent, showing no flat areas or overly sharp bends.

"I spent two days just developing a wooden jig that puts a perfect curve in the rods. I can now bend rods in seconds -- they use to take two hours apiece," states Mr. Harper.

The drive can be powered by either four AA batteries or 12 volt DC from a portable battery or a car's cigarette lighter. When AA batteries are used, the voltage goes directly to the main regulator. The 3 position power mode switch determines which power source is fed to the main regulator. Carbon-zinc batteries can run the drive for about 11 hours. The main regulator circuit has an input range of 2.5 to 6.5 volts DC. The output is an extremely stable nominal 5.00 volts DC.

The 12 volt DC is initially regulated to 5 volts DC before going to the main regulator. The circuit -- using either the AA batteries or a car battery -- is relatively immune to temperature changes. If batteries weaken from cold weather, the circuit will continue to output 5.00 volts DC.

Astro Kits is presently working on a more advanced electrical circuit that will provide a very stable output from the motor even if the camera/lens weight on the upper deck varies greatly. This compact circuit, which uses a stepping motor design, will soon be available to customers.

You can polar align this tracker by sighting down the very thin (1-mm) hinge line, first zeroing in on Polaris, then off-setting it 0.8 degrees towards the star Mizar in the Big Dipper.

Be sure to have the camera and lens mounted when you make the alignment; this will compensate for any slight sagging caused by the weight of the camera and lenses bearing down on the ball-and-socket head.

Use a finder scope, Mr. Harper suggests, for polar alignment whenever you shoot with lenses over 135 mm. More precision is required when these longer telephotos are employed.

An optional high quality metal hinge is available when longer, heftier lenses from 200 mm to 500 mm are used. According to Mr. Harper, Astro Kits standard polyolefin one-piece plastic hinge will work fine with lenses up to 135 mm in focal length.

Any disappointed customers? "Only one tracker has been returned," Harper says. "A photographer in Texas insisted -- against my warnings -- on using a 9 pound lens atop one of the plastic hinge models. Of course it didn't work. I'm replacing that hinge with a metal one and will send it back."

The cost:

Price of the complete kit:	$99.00
UPS shipping cost:	3.50

Total cost:	$102.50

Fully assembled and tested drive: $127.50

To order or to get more information:

ASTRO KITS
Dennis Harper, Owner
824 Bronx River Road
Bronxville, N.Y. 10708
Phone (914) 776-1931

APPENDIX E

SUPPLIERS OF STAR TRACKERS, MATERIALS OR EQUIPMENT

ASTRO KITS
824 Bronx River Rd.
Bronxville, N.Y. 10708
(914) 776-1931, 237-9970
(Tracking kits and parts)

CELESTRON INTERNATIONAL
P.O. Box 3578
2835 Columbia Street
Torrance, CA 90503
(213) 328-9560
(Stelas drive motor, finderscopes, batteries, telescopes and parts)

EDMUND SCIENTIFIC
101 E. Gloucester Pike
Barrington, N.J. 08007
(609) 573-6257
(Small DC motors, optics, telescopes)

A. JAEGERS
691 S. Merrick Road
Lynbrook, N.Y. 11563
(Finderscopes, optics, drive systems)

KAJERS
1112 W. Vine St.
Springfield, Ill. 62704
(217) 787-0861
(Camera tracker, drive systems)

LUMICON
2111 Research Dr. #5
Livermore, CA 94550
(415) 447-9570
(Hypered film, filters and lenses for deep-sky shooting, telescope accessories)

MEADE INSTRUMENTS CORPORATION
1675 Toronto Way
Costa Mesa, CA 92626
(714) 556-2291
(Finderscopes, telescope accessories)

OPTICA B/C COMPANY
4100 MacArthur Blvd.
Oakland, CA 94619
(Books, astrophotography supplies, telescope accessories)

ORION TELESCOPE CENTER
P.O. Box 1158
Santa Cruz, CA 95062
(800) 447-1001
(Finderscopes, telescopes, accessories)

PURUS -- EVERT ACHTERBERG
P.O. Box 123
Escalon, CA 95320
(209) 838-1259
(Clock-drive tracker)

TEXAS NAUTICAL COMPANY
2129 Westheimer
Houston, Texas 77098
(713) 529-3551, 529-8480
(The Takahashi drives and camera tracker, telescopes, finderscopes)

ROGER W. TUTHILL, INC.
11 Tanglewood Lane, Dept. HB
Mountainside, NJ 07092
(800) 223-1063
(Clock drive, optics, telescope parts)

UNIVERSITY OPTICS
P.O. Box 1205
Ann Arbor, MI 48106
(313) 665-3575
(Hypered film, optics, telescope parts)

BOOKS

ASTROPHOTOGRAPHY by Barry Gordon, published by:
Willmann-Bell, Inc., 1985
P.O. Box 3125
Richmond, Virginia 23235
(Mostly guided photography info; great reference book)

ASTROPHOTOGRAPHY FOR THE AMATEUR
by Michael Covington, published by:
Press Syndicate of the U. of Cambridge, 1985
U.S. address: 32 East 57th Street,
New York, N.Y. 10022
(Best available astrophotography book for the amateur. Worth reading)

BUILD YOUR OWN TELESCOPE by Richard Berry,
published by:
AstroMedia Publications, 1985
1027 North Seventh St.
Milwaukee, WI 53233
(Step-by-step plans for building five different types of telescopes; you may be able to use mount designs shown)

BURNHAM'S CELESTIAL HANDBOOK, 3 VOLUME SET
by Robert Burnham, Jr., published by:
Dover Publications, Inc., 1978
31 East 2nd Street
Mineola, N.Y. 11501
(Detailed, interesting material on stars; hundreds of photos, charts, graphs. Great reference set)

A FIELD GUIDE TO THE STARS AND PLANETS, 2nd edition, by Donald H. Menzel and Jay M. Pasachoff, published by: Houghton Mifflin, 1983.
(Includes whole-sky maps of constellations visible in northern and southern hemispheres each month; excellent maps by Tirion)

SKY ATLAS 2000.0 by Wil Tirion, published by:
Sky Publishing Corp., 1982
(Shows stars to eighth magnitude and numerous deep-sky objects, including diffuse nebulae)

SKYSHOOTING by R. Newton Mayall and Margaret W. Mayall, published by: Dover Publications, Inc., 1968
(Fascinating book about early astrophotography equipment and techniques)

THE STARS by H. A. Rey, published by:
Houghton Mifflin Company, 1980
2 Park Street
Boston, Mass. 02108
(A classic; fine guidebook for beginners)

365 STARRY NIGHTS by Chet Raymo, published by:
Prentice-Hall, 1982
Englewood Cliffs, New Jersey
(More detailed than THE STARS, this book has excellent material about stars)

MAGAZINES

THE ASTROGRAPH -- BIMONTHLY --
P. O. Box 2283
Arlington, VA 22202
(Articles on astrophotography techniques; mostly piggyback shots from telescopes)

ASTRONOMY / DEEP SKY MAGAZINE -- MONTHLY --
AstroMedia Corp.
1027 N. Seventh St.
Milwaukee, Wisc. 53233
(You should subscribe to ASTRONOMY)

MERCURY -- BIMONTHLY --
Astronomical Society of the Pacific,
1290 24th Ave.
San Francisco, CA 94122
(Some astrophotography articles; provides helpful reviews of current astronomy books)

SKY & TELESCOPE -- MONTHLY --
Sky Publishing Corp.
49 Bay State Road
Cambridge, Mass. 02138
(You should subscribe to SKY & TELESCOPE)

TELESCOPE MAKING -- QUARTERLY --
AstroMedia Corp.
1027 N. Seventh St.
Milwaukee, Wisc. 53233
(For do-it-yourselfers; often have good articles on customized mounts)

PERTINENT ARTICLES

Berry, Richard, "Astrophotography -- With Camera Only!", ASTRONOMY (Jun 78), 42-47.

Coppola, Mike, "A DC Motor Speed Controller", COMPUTERS & ELECTRONICS (Oct 83), 57-62.

di Cicco, Dennis, "Notes on Gas Hypersensitizing", SKY & TELESCOPE (Feb 81), 176-177.

di Cicco, Dennis, "A Simple Camera Mounting For Astrophotography", SKY & TELESCOPE (Oct 85), 391-393.

di Cicco, Dennis, "Shooting Stars", POPULAR PHOTOGRAPHY (Feb 86), 11-17.

Dilsizian, Rick, "Motorized Astrophotography Using A Motorized Camera Platform", THE ASTROGRAPH (June/July 87), 83-89.

Haig, G. Y., "A Simple Camera Mounting for Short Exposures", SKY & TELESCOPE (Apr 75), 263-266.

Haig, G. Y., "Equatorial Mount for a Camera", ASTRONOMY NOW, vol. 1, no. 1, (Spring 87).

Holbrook, John, "A Stepping Motor For Your Telescope", SKY & TELESCOPE (July 86), 80-83.

Hooley, Mace, "Fundamentals of Astrophotography", ASTRONOMY (May 86), 58-63.

Hooley, Mace, "Astrophotography with Tech Pan 2415 Film", ASTRONOMY (July 86), 76-79.

Hooley, Mace, "Processing Astroslides Yourself", ASTRONOMY (Sept 86), 102-105.

Hooley, Mace, "Advanced Astrophotography", ASTRONOMY (Nov 86), 50-55.

Hunter, Tim, "Improving Astrophotos by Copying", SKY & TELESCOPE (Sept 87), 326-328.

Mims, Forrest, "Controller for Small DC Motors", POPULAR ELECTRONICS (Oct 82), 98.

Pursell, Wallace, "A Portable Photographic Platform", ASTRONOMY (July 87), 91-93.

Trott, Dave, "The Double-Arm Barn-door Drive", SKY & TELESCOPE (Feb 88), 213-214.

Wahman, Joseph, "Shooting Stars", PETERSEN'S PHOTOGRAPHIC (Oct 85), 38-42.

Zullo, Frank, "Shooting Stars", OUTDOOR PHOTOGRAPHER (Jan/Feb 86), 55-59.

Zullo, Frank, "Shooting Stars", DARKROOM PHOTOGRAPHY (Nov 87), 20-27.

Your comments about this book are appreciated. Write and tell me about your tracker and how well it performs.

Send your letters to: Jim Ballard
10614 32nd Dr. S.E.
Everett, WA 98208

Please enclose a self-addressed, stamped envelope if you want material returned.

INDEX